CHITRAL SCOUTS

Illustrated History of Chitral Scouts 1903-2014
Major Aamir Mushtaq Cheema
First Published in 2014 by War Studies Pakistan
First Printed by The Army Press
Plot #1, Street 40, I&T Centre, G-10-4, Islamabad 44000, Pakistan
Reprinted in 2018 by Ishi Press International
1664 Davidson Avenue, Suite 1B
Bronx NY 10453
USA

Aamir Mushtaq Cheema <aamircheema@outlook.com>

ISBN 4-87187-181-9
978-4-87187-181-5
1-917-659-3397
1-917-507-7226
samhsloan@gmail.com
Printed in the United States of America

ISHI PRESS
INTERNATIONAL

An Illustrated History
of
Chitral Scouts
1903-2014

Table of Contents

Glossary

Adjutant: An officer of Captain or Major rank he is responsible for discipline, training, principal staff officer.

Commandant: An officer of Colonel rank, who commands the scout, he is from regular army.

F.C: Frontier Corps, Present day Frontier Corps Khyber Pakhtunkhwa.

Kashkar: Ancient name of Chitral.

S.M: Subedar Major, the highest rank which a soldier can attain.

J.C.O: Junior Commissioned Officer

M.I: Mounted Infantry

N.C.O: Non Commissioned Officer.

Officers: They are Commissioned, from regular army, trained in Pakistan Military Academy.

Q.M: Quartermaster, an officer of captain/major rank looks after the administration of scouts.

Qaum: Nation, it is the tribal concept of unity.

Wing: It is equivalent to an infantry battalion but is lightly equipped. Commanded by a Lieutenant Colonel known as wing commander, each wing is composed of four riffle companies.

Foreword

Colonel Naeem Iqbal
Commandant Chitral Scouts, Frontier Corps

'Chitral' with its distinctive culture, norms and social matrix has no parallels. Diverse religious values with apt absorption, hospitality and simplicity are hallmarks of this soil. Same parameters on consolidation, form onerous and honourable **'Chitral Scouts'**. History of Chitral has correlation, attribution and an inimitable binding relation with the Scouts' history. The outfit, from a modest beginning has come a long way, making each day momentous, adding elevation to our exalted history. Contemporarily, once tribal wars, intrigues are rising in rest of the province, Chitrals have stood singularly for patriotic fervour over a singular platform. The model, followed in Chitral, perhaps stands distinguished in solving an insoluble problem, we are badly infected with. Chitral Scouts being a solitary and diligent asset of Frontier Corps is a pride for locals, who trust us with absolute belief that we will always be with them for securing our borders, their culture and traditions, being custodian of all.

I bow myself in complete humility to Almighty Allah for providing me an opportunity to command this unique set up. On my arrival, I wanted to learn about the retrospective decades in order to bring further blossom to this cherished garden. Being determined to learn about broad story line, I found lots of information either mission or uncertain. I took a challenge to fill in the glaring gaps with accuracy and precision. But on various pretexts, was still short of applying analytical tools in collection methods, once I learnt about Maj (Retd) Aamir Cheema's efforts in compilation of Chitral Scouts history. Surprisingly, soon the book was with us for deliberation as a blue print. I feel gratitude in saying that Maj Cheema has put in a intellectually rigorous, systematic and profoundly honest effort in compiling the accounts. His sagacity and commitment has left a remarkable print on the Unit and his effort will see annals of history. The book is a singular tribute nothing but memorable and eternal, being an account of truth. May this book go a long way in educating many on Chitral and Chitral Scouts.

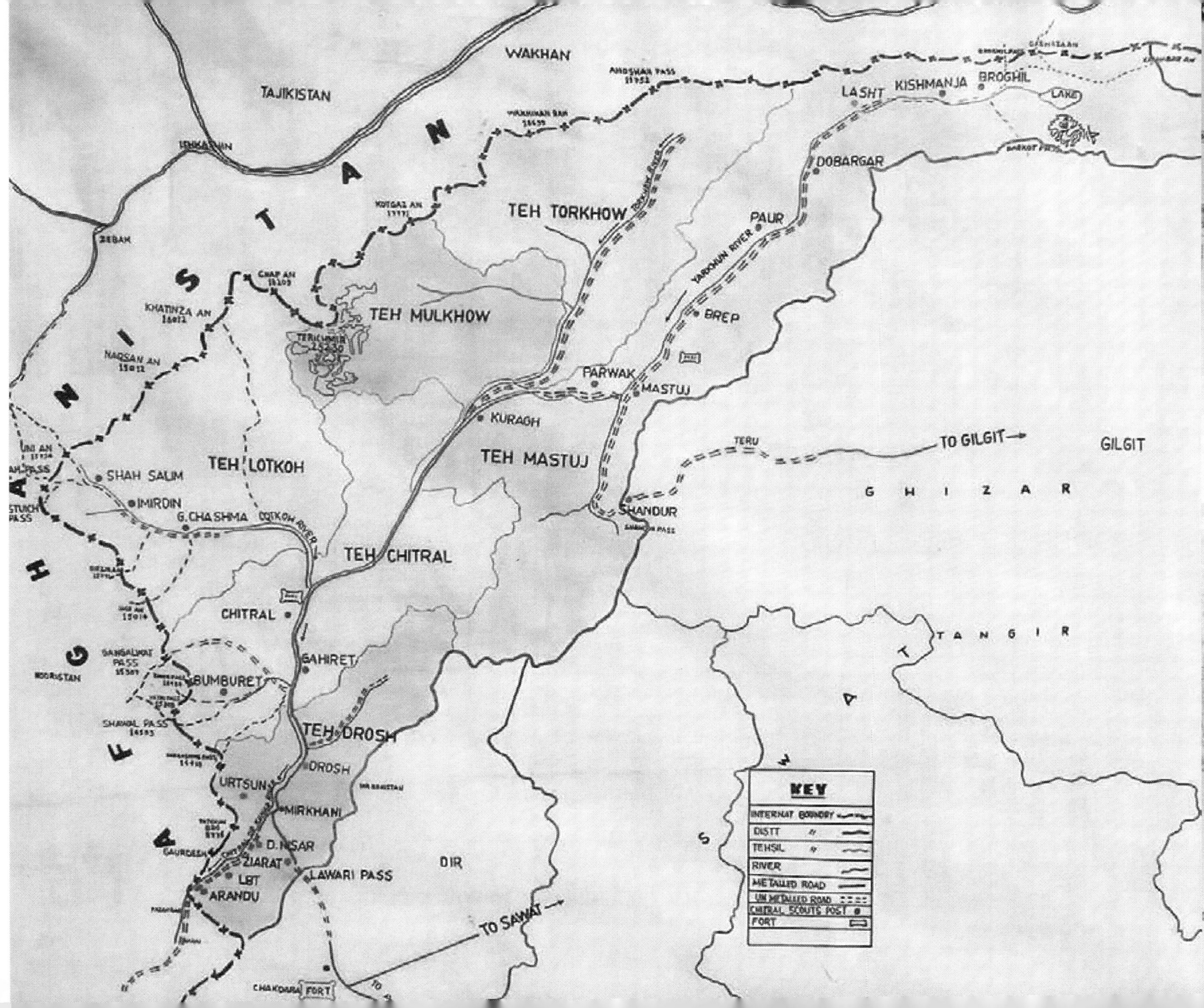
WAKHAN
TAJIKISTAN
A F G H A N I S T A N
LASHT
KISHMANJA
BROGHIL
LAKE
DOBARGAR
TEH TORKHOW
PAUR
YARKHUN RIVER
TEH MULKHOW
BREP
PARWAK
MASTUJ
KURAGH
TEH MASTUJ
TEH LOTKOH
SHAH SALIM
IMIRDIN
G.CHASHMA
SHANDUR
TERU
TO GILGIT
GILGIT
G H I Z A R
TEH CHITRAL
CHITRAL
SAHIRET
BUMBURET
TEH DROSH
DROSH
URTSUN
MIRKHANI
D.NSAR
ZIARAT
LBT
ARANDU
LAWARI PASS
DIR
TO SAWAT
CHAKDARA FORT
T A N G I R
S W A T
KEY
INTERNAT BOUNDRY
DISTT
TEHSIL
RIVER
METALLED ROAD
UN METALLED ROAD
CHITRAL SCOUTS POST
FORT
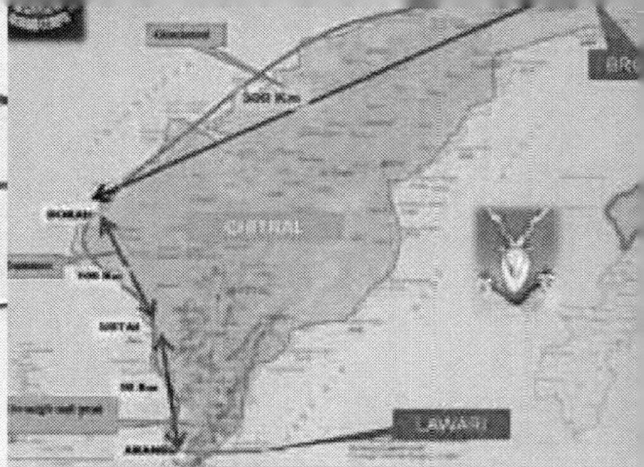
CHITRAL
LAWARI

اسلامی جمہوریہ پاکستان
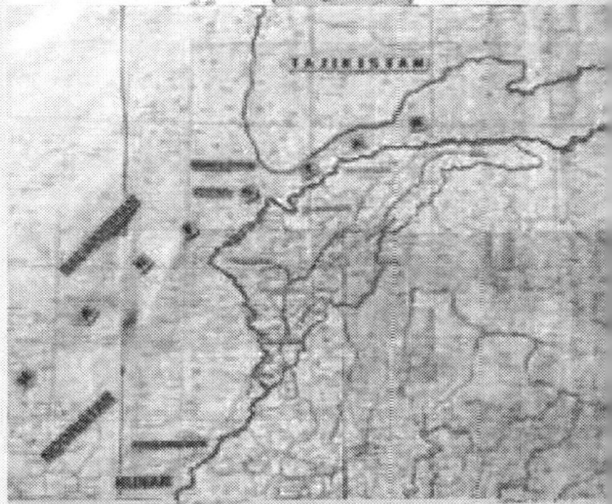
TAJIKISTAN

ضلع چترال

Chapter One

Chitral Scouts & Chitral State Scouts 1903-2014

Chitral Scouts were raised on 23rd July 1903 and on the eve of First Great War it had a strength of almost a thousand men organized into two wings. In 1942 it was named as **Chitral State Scouts** and came under the control of Frontier Corps, in 1953 it reverted to its old nomenclature. Chitral State Scouts played a key and vital role in the Kashmir Liberation of 1947-48. In June 1973 another wing was raised, followed by another one in June 1986 ; and next year in July two more wings and an artillery battery were also added. It was raised at Chitral in 1903 , shifted to Drosh in 1926 and later moved back to Chitral in 1992.

Chitral District covers an area of 5000 square miles and had a population of 315,000 in 2001. In 1895 the population was under 1,00,000 humans, and 1,17,000 in 1961. Chitral is a confluence of two valleys and rivers , the western being Tirch Mir Valley which takes origin from 25230 feet highest peak and is 40 miles long with Garm Chashma as the major town and Shah Salim Pass as the doorway. The eastern Yarkhun Valley runs a distance of 250 miles starting from Sor Yarkun near Oxus few miles east of Broghul Pass in the extreme north to the Chitral, in south Mastuj and Booni are the major towns. There are more than 40 peaks above 20,000 feet in the district. The two rivers join together near Chitral and from hence known as River Chitral and after exiting the district boundaries at Arandu it is known as River Kunhar.

Chitral takes its name from the Chitral town which is situated almost in the lower middle of the district, it attained the status of a settled district from a princely state in August 1969; in April 1972 all titles were also abolished. Pesently Chitral is divided into two administrative layers, the Upper Chitral is known as Mastuj sub division and comprises of three tehsils namely Mastuj, Torkhow and Mulkhow with Booni as the headquarters, Lower Chitral sub division comprises of Chitral, Drosh and Lotkow tehsils. History of the Chitral Scouts is mainly spread in the Lower Chitral areas. Upper Chitral is also known as Chitral Bala and lower Chitral as Chitral Payan, Chitrar and Kashkar are other names of Chitral.

Chitral possess a unique feature in the present day world history, the presence of 3000 odd Kafir tribesmen among over whelming peaceful Muslims; there are no Christians, Hindus or Sikhs in Chitral. The Kafir tribe is settled in the lower Chitral on the western bank of Chitral River scattered in three valleys namely the Rumbir, Birir and Bhumbirit. Physically, Chitral is the largest district of Pakistan in terms of area. Chitral is like many other mountainous towns, a river passing through it, green valley, houses on the top, extreme winter and peace. Unlike Tochi Valley it is much more green and wide with higher peaks. As compare to the Kurram Valley it is narrower with more steep banks and field elevation. Gilgit and Chitral seems to be identical in nature and their history is also similar in pattern. It is the presence of colourful Kafirs which gives Chitral its unique flavour of history.

Chitral Scouts have quite a few distinctive marks, one they do not speak Pashtu secondly they are from a settled district which was a princely state in the past and above all they all are from one qaum. There are other cultural features as well, geographically they remained cut off from rest of the

Western Chitral as seen from Chitral Scouts Officers Mess

country during winter months ,; their only access via Afghanistan . Not long ago Toyota Hiace used to ply between Peshawar and Chitral during winter through Jalalabad into Asmar then into Arandu Valley. Lowari tunnel is still not open for public use but in winter traffic is allowed thus a millennium old issue has been resolved. Lowari Tunnel project was initiated by the Prime Minister Zulfiqar Bhutto in 1973 and completed by President Musharraf in 2007.

A Convoluted History of Chitral {Kashkar}. Chapter One.

Chitral's history is interlinked with the history and geography of Dir, Swat, Gilgit and Kashmir apart from Kabul. In other words The Tirch Mir, Nanga Parbat, Rakaposhi, K-2 are inter connected, one can view the same in terms of mehtar, nawab, khan, rais, maharaja, thum and political agent. It is also a conflict & cooperation between the Hindukush, Karakorum and Himalayas. What is written about Chitral is mainly by the Chitralis which primarily revolves around the prince's or royal history; something very similar to England.; Military history of Chitral is almost extinct.

Chitral has a history which is interesting, colourful having all segments of life, although it is not yet fully understood. The state known, as Chitral in perception is an amalgamation of many small fiefdoms, serfdoms and odd tribes' most obvious being the Kafir. Chitral is in fact a combination of two major riparian cultures and civilisation namely the Tirch Mir River and Yarkhun River; both meet just a mile upstream of present day airport and just miles away from the century old fort and the city itself. These two rivers express two distinct identities and cultures and it is only at Chitral that these two cultures join together. River Gilgit is also part of this specific mountainous entity.

TirchMir was the mountain god standing majestically at the west of Chitral, the early morning sun rays casting golden radiant around it and at sunset the crimson color bidding farewell to the day's journey. Aryans worshipped mountains, they were remote, inaccessible yet visible with efforts. Clouds around them bringing rain and darkness. At night these clouds would wrap the full moon around their web creating myths and fables; Tirch Mir is no exception rather it is among those very rare mountains along with Rakaposhi and NangaParbat to have fertile valley running around them. Even among these Tirch Mir stands out as it is more fertile, plain and green in nature. West of Chitral town is the confluence of two rivers which are also two distinct paths of history. The eastern River Yarkhun leads to further north into Booni, Sonoghar, Mastuj, Wakhan , Kashgar and into China; { or towards the Rakaposhi, K-2 , }, where as the western or TirchMir river leads into Afghanistan{Badakshan}.

Aryan migration started in 2000-3000 years BC from Central Asian highlands which were climatically extreme in winter with little to grow moreover the built in natural instinct of humans must have drove them downwards.; thus there is nothing abnormal to write that the early civilisation took birth at the Tirch Mir and from there it moved down ward and towards the west with the flow of water. Aryans not necessarily were violence lovers, among them there were some, having temper bit higher than others, it seems the mild tempered and intelligent among them settled among these upper nature blessed areas. These areas are nothing short of heaven, fresh water both cold and warm year around, streams full of fish, fruit of all kind, apple, grapes, walnut, apricot, peach, pomegranate, wild birds, sanctuary of migrating birds, ibex, Markhoor, snow leopards, bear, trees that few can match their grandeur like maple and pine covers it like a shawl; list can go on. Mountains have a charm of themselves and in our entire mythology world over it has always been the mountains which remains the abode of god, lest it not be forgotten that Moses{pbuh} went up the Mount Sinai to have the commandments.

Tirch Mir was the first of the many valleys or waterways which the Aryans adopted for their migration downwards. The beauty, serenity, peace and above all the abundance of fresh water all along must have been very tempting for them to settle astride the water flow; Chitral's history is interwoven with other mountainous states. Being on the edge of all of them it attained a greater strategic significance. It is the door way to the Central Asia or vice versa a route to the plains. On the north it is bounded by Oxus or the Wakhan, on the west by Gilgit, Yasin and Ishkoman mountainous states, on the south by Dir and Swat and in the south west by the Afghan province of Asmar and Nuristan.

Alexander the Great 323 B.C. came down towards Indus from the Oxus marching along the fringes of Tirch Mir and there are all the reasons to believe that he crossed into the Tirch Mir River culture and flowing with it marched through the Ayun, Drosh, Mirkhani, Dommel Nissar, Arandu and pouring into Bajaur. Chitral is not mentioned by Herodotus or by the Alexander the Great's historians. Present Chitral as such never existed in ancient times, it was known as Kashkar or Kashkaro[1]. Basing upon the geography it is logical to assume that majority of his troops must have intruded into the Chitral Valley as no less than 19 mountain passes were available to Alexander while marching down from Oxus towards Indus. It must be kept in background that Alexander's history was written almost 400 hundred years after his death by Arian while sitting at Egypt. There are confusion in names, places and dates. Greek mythology revolves around Dionysus the last of Olympus and Tirch Mir full fills the description, after all Dionysus is related to grapes & wine making[2]. Moreover the Panjkora River which takes origin from the Kashkar's{Chitral} eastern wall 'Hindu Raj' is very much covered by Arian in his history and also by host of other earlier Greek historians who all unfortunately never visited the area to ascertain the fiction from facts basing upon geography.

Buddhist Chitral

599-527 B.C., by this period some of the Aryans had been assimilated in the life of a settler, they now live in cities which became small kingdoms spread all over the India , Meghda was one such kingdom.

In the same time period a son was born to Suddhodana king of Magadha, he named his son as Sakyamuni Gautama he was popularly known as Siddhartha and today is famous as Buddha[3]. Magadha was one of the leading and rather the most powerful kingdom of the eastern India which existed at that time it comprehends all the Gangetic provinces[4]. Thus it seems that whereas the area astride the Indus was under the Persian as their province the area further east i.e. the Ganges plains were under the descendants of the Aryans from where a new social system was taking birth in the form of Buddhism.

For twenty seven years his father shielded his son from the cruelty of the world, the death, the old age, sickness and misery but one day Siddhartha was able to see it when he went outside his palace. Buddha was married at the age of nineteen and had a son also, for six years after his first encounter with the dark side of the life he wondered around and mediated and fasted more or less in the pattern of the Mahavira , he finally attained nirvana under a tree.[5]

1 Munphool Pundit, on *Chitral, Gilgit and Kashmir.*, p-38.
2 Metropolitan Museum of Art New York City; plaque placed in hall No.1. October 2011.
3 *A Buddhist Reading* , Dough Vincent, approved but unpublished Thesis, University of Western Ontario,2007 also see Buddha &His Doctrines, a bibliographical essay, Otto Kistner,{London,Trubnar & Company,1869}
4 *Researches into the Physical History of Mankind*, James Cowles Prichard, Third Edition, Volume IV, [London 1844] p 113.
5 *History of Asia*,Volume I ,Pp-143-145

River Injiman or Tirch Mir River

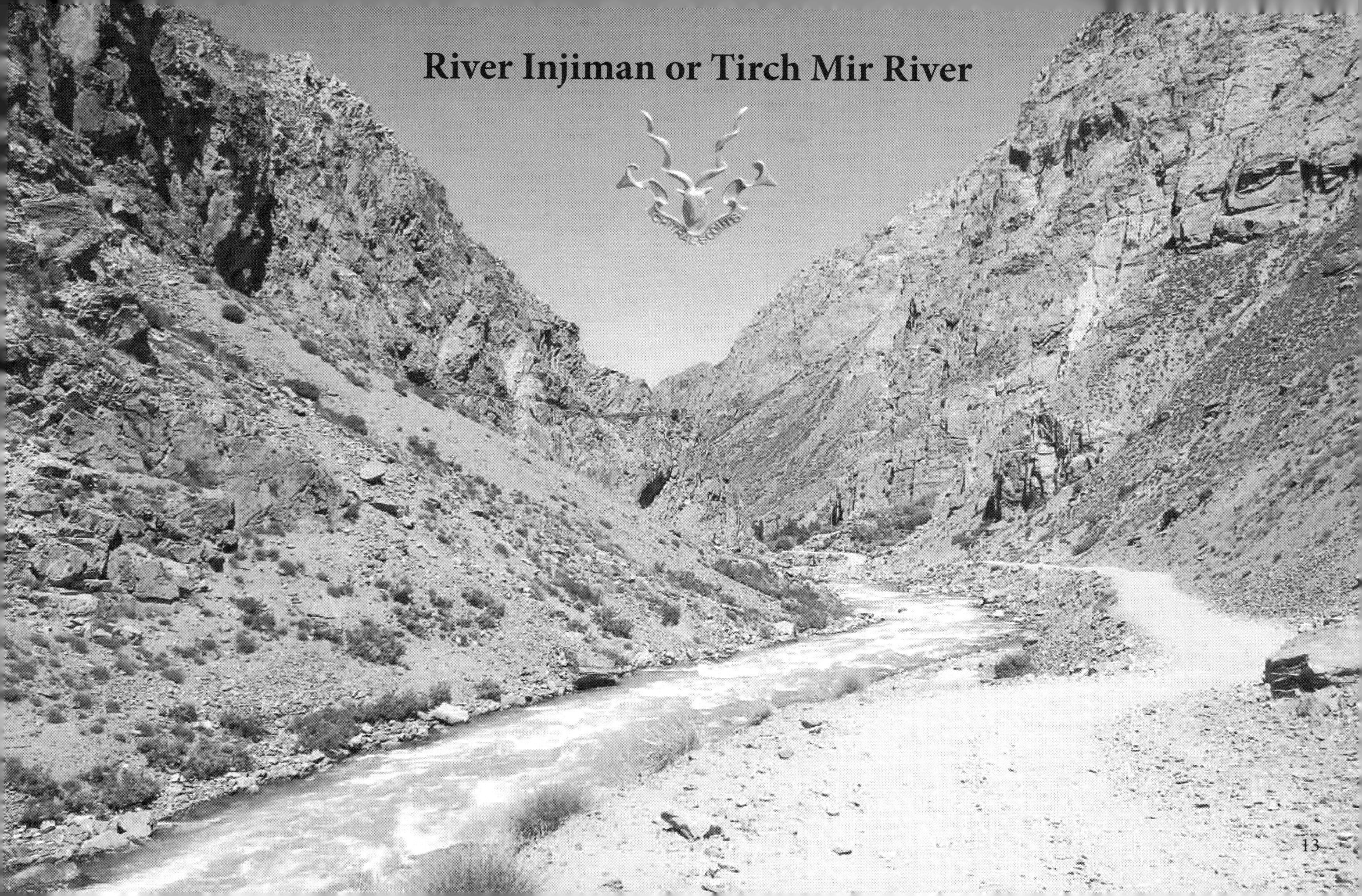

Buddha was a revolutionary probably the first socialist of the world, he was against the caste and promoted a classless society free of rituals and rites where all were equal, Buddha died in 543 B.C., and it was not almost 235 years after his death that his teachings were collected and they continued to be collected under the reign of Asoka in 200 B.C., in the year 420-500 C.E. there existed a great Buddhist university at Nalinda. However later the Bhuddism was divided into two main sects one is called Hinayana which does not believe in his divinity and neither worship him in the form of images the other Mahayana which makes images of Buddha and considered him divine, it is Mahayana which spreads in China and in India thus the engravings of Buddha in Mastuj {upper Chitral} Hunza, Gilgit, Skardu. This also explains the origins of Kafirs, they probably came from east of Maghda to Chitral with what was probably the very first 'Hippy Trail'.

History on Metal, Chitral Scouts Officers Mess

The period from Ashoka till the arrival of Islam in 700 AD is termed as the dark ages of India by historians, for the reason that very little historical account has been found of that period. Tirch Mir Kingdom was under the suzerainty of King of Kabul, Raja Jaipal. It was an Aryan kingdom, Hinduism itself cannot be expressed in the same language as one can use for Christianity; it is a conglomeration of over five hundred different sects ranging from worshippers of Phallus{Shiva} to the Monkey {Hanuman}. Chitral was one of the finest among such fanatics. It had very less population probably not more than twenty odd thousand humans and equal number of animals; it was certainly a popular spot for migratory birds coming from Siberia as they still land here. Chitral then was ruled by the Sia Posh or Black Kafirs. The upper valleys or Upper Chitral was divided into many smaller Khowistans, Torkhow the upper Khow, Mulkhow the lower Khow these names persists even today. Buddhism had its impact in the area with one odd inscription of 9th Century AD records the fact that the area is under King Jaipal of Kabul. Khowar language has quite a number of words from Sanskrit in it. One stupa of Buddha is still intact at Torkhow known as Kalandar-i-Bhuttani, {mendicant of stone}.

Chitral also remained a well reputed spot for flesh market; Badakshan across the valley in Afghanistan was famous for its horses. This can be one explanation of such a variety of languages as caravans moving west towards the Kashmir preferred travelling through Arandu- Drosh- Madaglasht- Tangir- Chilas crossing River Indus; thus having only one river to cross instead of two if they choose any pass west of Chitral. Babur the Great had to cross three rivers every time he came down upon India. On the other hand , caravans moving from Kabul towards Eastern Turkestan{Sinkiang} also preferred Arandu Pass. But by and large Chitral remain hidden in history and present hypothesis about its history are based upon geography which have remained intact in all these millenniums.

Islam knocked on the Chitral or Khowistan's door in the aftermath of the Turk's defeat in the hands of Arabs in 751 AD[6]. Broghul Pass has been the main entry route of Mongols and Tartars in the past. Marco Polo referred to the country as Bolor which included Gilgit as well. Another strand of history recalls a Chinese tribe by the name of Yarkhun to have invaded the area. **By 10th Century AD** the area was ruled by the kafirs however Mahmud of Ghazna's rise and his utter destruction of idols in the India must have been the cause of the downfall of kafirs and slowly but gradually they started getting into the small valleys. Tamerlane's conquest of India and his slaughter of Hindus was the last of the nails in the Kafir culture. The area however retained its identity as it even today is known as Kafiristan although the major portion which lies in Afghanistan is now known as Nuristan.

Engraved History, Chitral Scouts Officers Mess

Chagtai Khanate-Marwa Ul Nihar

Great Khan 'Temujin' had divided his empire which at time comprised of China, Central Asia, Middle East, Afghanistan among his three sons. Present day Afghanistan, Pakistan, Tajikistan, Uzbekistan and Chitral were all part of Chagtai Khanate. Mughuls are Aryans, they belong to Turk ethnic group and they were settled in the Central Asia or more precisely Eastern Turkestan which comprised of Samarkand, Bukhara, Fergana, Tashkent, Kashgir cities. They were in area between River Oxus {Amu Darya} and River Jaxarates{Sur Darya}thus only Amu Darya and Wakhan Strip separates them from Chitral , Hunza, Gilgit onwards to River Indus. Chitral has striking resemblance with the vales of Fergana, the birth place of Mughul dynasty which ruled India for three hundred years. It was Amir Taimour {Tamerlane} who first ventured into the adjoining areas during his initial days in power. Chitral like Fergana was composed of several small valleys without any strong central authority, poor in resources and harsh climatic conditions.

The Eastern Chagtai Khanate also known as Marwa Ul Nihar[7] was Muslim by faith following Hanfai & Malaki sect , which is more liberal in interpretation of Islamic laws for instance almost all Mughul emperors were fond of liquor yet detested pig meat on religious grounds and even in dire hunger never took it[8]. They

6 S.A. Malik *The Muslim Conquest of Central Asia,*{ Army Education Press, Rawalpindi, 1979}. Aurel Stein, *The Geographical Journal Volume,59, No.2 {February 1922}* A Chinese expedition across the Pamirs and Hindukush, A.D. 747.pp112-131.

7 *Tuzk-i-Taimouri,*Autobiography of Tamerlane translated from Persian by Syed Abu al Hashim Nadvi {Sange Meel publishers, Lahore, 2012}p,21.

8 *Tuzk i Babri,* Autobiography of Emperor Babur, translated from Persian into Urdu & English., p-9.

would pray and give alms, keeps fast in the holy month of Ramadan and sacrifice lambs in Zi ul Haj month. Family life was important to them and they seldom left their wives for long, always respectful of in laws. They were gentlemen in nature, adhering and following a code of warfare based upon the Temujin's Code and that of Prophet Muhammad {pbuh} directives. One of the fundamental aspect of Islamic warfare is deception 'War is nothing but deception' and Timurids followed it by spirit. Assassination was a common method to remove the leader by any means; poison was most favoured method. Battles were fierce and had to be fought with a character; the life of the vanquished depends upon his conduct in the battle he had recently lost. The looser had to put the grass in his mouth to show his surrender. Courtesies were immense, a khan or mirza would greet the equivalent by receiving him almost two miles ahead of his camp. Headgear was important and none was seen without it, khans would send the caps as mark of respect and gift to fellow family members. Food was in shape of meat and fruits, all this holds true for Chitral as well. Yet the great wall of Tirch Mir kept Chitral out of the main focus, there are no individuals in the art of warfare worth mentioning.

Babur after capturing Kabul learnt about the Kafiristan but took no immediate action. In the power tussle that was taking place in Fergana the small states like Chitral had no major role to play but the loyalists of Babur started plotting in these small states and when and if succeeded were rewarded with the kingship of the area; similar story took place in Chitral. In 16th century the Chitral River banks were under the control of Mughuls with Shah Rais on the throne. His advisor was Sangin Ali a Persian, with the passage of time the family of Sangin Ali came into power and remained so till the end of the state. Within the family the similar pattern of in fight as seen in Mughuls followed which resulted in lineages; the Katoras take their name from the Sangin Ali's grandson where as the Khushwaqts take it from the second son. Historically there were two major ruling dynasties the rais which were ruling Chitral from 1000 -1530 AD and Kator, these two ruling classes have biter and bloody history trail like all other princely houses all over India or for that matter in world; it still persists.

Dardistan, Dagestan, Kohistan, Kafiristan & Khowistan are few of the geographical identities which have been associated with the east and west Hindukush areas; among all these Khowistan {Chitral} was the only one which was named due to the language bondage**.** The area east of Chitral encompasses Kashmir, Swat, Dir and East Turkistan and interaction among them in last two hundred years have set the stage for present and to some extent the future of the area. Geographically the area between the River Gilgit and River Chitral can be taken as one entity and it was one entity as late as 1850 when the Gilgit was under the Mehtar[9] Aman ul Mulk . Swat was not a state as yet then, Nawab of Dir and Khan of Jandol were two other strong contenders of the suzerainty but the Hindu Raj mountains kept them at bay . It was the Arandu Pass which remains open year around and provides an easy access to the Pathan and Afghan to venture into Chitral Lower. Upper Chitral remained almost independent even during the reign of Aman uk Mulk, it had the visitors from the west passing over the Broghul Pass and settling into the villages of Tor Khow and Mulkhow. The Lotkow the western area of Chitral had the Afghan or Tajik travellers coming across the mountain ranges via Shah Salim Pass. In nut shell, Chitral's passes provides the gateway to the northern parts of India, Kashmir and now Pakistan. Broghul can be rightly called as the door to the Kashmir. Even now the population of Broghul hops over the Darkot Pass to have the supplies bought from Gilgit traders rather than coming to Mastuj or Chitral, because the distance involved is quite less at Darkot.

The Upper Chitral is predominantly Ismaili by faith and Lower Chitral is Sunni. Ismaili is a sub sect of Islam which believe in the living Imamate, Prince Karim Agha Khan is the current imam. In the making of Pakistan, Prince Karim's grandfather Agha Khan played a key and vital role , in fact he headed the Muslim League for almost a decade and half since its creation in 1906. Traditionally it has always been the eldest son of

9 Imperial Gazetteer Provincial Series North West Frontier Province, 1st edition 1910 (Sang-e-Meel, Lahore 2002).

living Imam who is nominated but it in case of present imam it was not followed and instead of his father Prince Aly Khan ; he was nominated thus a section within the Ismailis also exist. Upper Chitral has influence of Tajikistan, Badakshan and Pamir where as the lower Chitral is under the influence of Pathan culture or Kunar culture which creeps up stream through the Chitral River. Socially the society comprises of the ruling families known as Adamzads, Arbabzads come next in ladder followed by the Fakir & Maskin{ same as in Aryans}. There had been religious persecutions; as late as in 1926 Mehtar carried out an all out effort to convert the Ismaili's living in Upper Chitral into Sunnis , it happened after his hajj. The persecution was stopped only after the intervention of British political officials.

Present History

Present history of Chitral begins in 1819 when Maharaja Ranjit Singh captured Kashmir and handed it over to the Dogras as a reward and later British after the defeat of Khalsa in the Anglo –Sikh wars of 1842-1848 sold it to the same Dogra family . Kashmir under the Ghulab Singh expanded outwards to the territories hence alien to the natives of India. They moved east ward towards the Tibet, northwards towards the Laddakh and west wards towards the Gilgit across the River Indus. This was the first and till to date the last invading force crossing Indus into Karakoram mountain range. By 1842 Kashmir durbar had a foothold at Gilgit and became the king makers, a garrison was established at Gilgit as well. The extreme northern states like Hunza and Nagar remained free from these incursions. In the present day Swat there was no state rather a loose conglomeration of khanates like Tanger, Darrel which were rather big in size and located on the western banks of River Indus. Kabul was the other major stake holder having a claim to the area of Kashmir as well, thus these small mountainous hamlet states became a battle prize among these two predominant states. Sikhs were religiously tolerant of Muslims and they had the big advantage in having a beard and a turban as part of religion both these items are integral part of western bank of River Indus's cultural heritage and religion.

After the annexation of Punjab in 1849 by the British East India Company {BEIC}, Chitral at that time was ruled by the Kator , there were two distinct families based upon geography who were the ruling elite, the Khuswaqts in the upper Chitral with stronghold at Mastuj and the Kators who were living in the lower Chitral. It was a classical narrative as both had the common blood running in their veins but both trying to extend their hold. Numerous small principalities in the small valleys put their weight behind each family basing upon their chances of success. Murder was common feature to grab the power. Maharaja of Kashmir made a move and had an alliance with the Kator Shah Afzal; Mehtar of Chitral in 1854 against the Khuswaqt Gohar Rehman of Mastuj who was attempting to attack Gilgit with the aid of Yasin's ruler. This was the start of Kashmir getting involved into the Chitral's affairs. In 1855 for a short period Chitralis occupied Mastuj but were driven out. On the other hand Shah Afzal Kator died in 1856 and after an intriguing episode his son Aman Ul Mulk became the Mehtar in 1857 and ruled till 1892. He is the father of modern day Chitral. The Khuswaqt of Mastuj had Gilgit under his control from 1848 till 1860 when Sikhs fought it back , Mehtar Aman Ul Mulk tried to keep the Gilgit under his control by force after the death of Gohar Aman but failed. However the Yarkhun Valley successfully repulsed an attempt by the Mir Muhammad Shah ruler of Badakshan in 1868 to annex the area; a decisive and bloody battle was fought at Durband.

Aman-ul-Mulk to unite the upper and lower Chitral, expanded his territory which included Gilgit, Ghizar & Thui and Chiga Sarai in present day Afghanistan. Khuswaqts of Mastuj on the other hand remained independent led by Khushwaqt Pahlwan, they made a venture to capture Gilgit in 1880 but the military weakness of Mastuj was once again exposed; in his absence Aman Ul Mulk captured Mastuj and that ends the story of Khuswaqts. Aman Ul Mulk later more importantly opened up the country for the British.

British in North. British venture into the north started in 1830 when travellers and foreign department officers disguised as saints and travellers gathered knowledge about the remote north. Names like George Hayward who was murdered in 1870 at Darkot, Mason , Burnes of Bukhara and later George Robertson who was the first British to live in Kafir territory made headlines. Britain established an agency at Gilgit in 1880 with Colonel Biddulph as the political agent but it was short lived. In 1885 the need to understand the western approaches to India were felt. The threat of Russia making an incursion through the Broghul Pass was a distant reality. An army team led by Colonel Lockhart visited Chitral and met Mehtar Aman Ul Mulk.

In 1892 the Gilgit Agency was again established which also monitored the affairs in Chitral. Chitral after Colonel Lockhart was visited by Colonel Algernon Durand in 1888 and 1889. Chitral or more precisely Aman Ul Mulk played power politics, in the past Chitral had been a subject of Badakshan in a nominal manner and now Kabul asserted the same pressure. Mehtar tried to have a betrothal with Kabul but it did not take place, on the other hand when Mehtar felt threatened from Kabul he made an overture towards Kashmir in 1874 which resulted in a kind of pact between two sates linking the defence of Chitral with Kashmir; British were the referee; in the end Chitral came under British sphere of influence and this became a bone of contention between the Kabul and Calcutta.

1891 was another water shed year in the history of northern areas. The Nilt and Hunza valley rulers became embroiled in a clash with the British led by Colonel Durand with British and Kashmir troops under his command. In December the forts at Nilt were overcome after great dexterity shown by the troops. Chitral remained neutral in this affair; Gilgit Scouts took birth after this expedition and a permanent agency was established at Gilgit with political agent in charge. It must be kept in mind that agency was established at Gilgit but it still remained under the Kashmir control. It was only in 1936 that the agency was leased by the British from Kashmir.

Princely States of India

Chitral , Dir, Swat, Amb, Kalat were princely states , there were very few such states which became part of Pakistan and where Frontier Corps was employed or which became part of it thus it is pertinent to have a brief look at the overall political system of India. There were two terms , one 'India' and other 'British India' to denote the present day Pakistan, India and Bangladesh before 1947. British India encompasses all the area which was previously under the British East India Company and after 1857 came under the jurisdiction of Queen Victoria. British parliament appointed a viceroy, usually for a tenure of six-eight years. In India the viceroy was the last word on all affairs, he had a commander in chief of army, a secretary of foreign affairs and other officials. India was divided administratively into provinces which included Bombay{Sind was part of it} Bengal, United

Provinces and after 1849 the Punjab; within which the NWFP was created in 1901[10] but it was not at par with other provinces as it was directly placed under the viceroy due to its strategic position.

Apart from provinces there were over 560 princely states of various size and population , the big states included Hyderabad Deccan, Jammu & Kashmir, Gwalior and Baroda to name few. British gradually moved westward and in this process certain states were annexed like Oudh in 1852 which became a province and many other were left at their own on certain conditions the paramount being , not to indulge or support anti British military campaigns. These states were ruled by princes, although all the states had different titles like raja, rao, nawab,thakur, nizam to name few but British classified them all as princes. Thus ' **British India'** denotes the area which was under British administration in form of provinces and **'India'** represents the princely states. 122 states were directly ruled by the central government from Calcutta {Delhi became capital after 1911} in other over 400 states there were political agents who were responsible to the provincial governors, commissioners and deputy commissioners. The princes of these states were given subsidy and were allowed certain gun salutes depending upon their importance. Most of the princes were entitled nine gun salutes. States were allowed to have a militia or army with British officers overlooking the affairs. Honorary ranks of lieutenant to general {very rarely} were bestowed upon the princes, states had their own flags and were not bound to follow the British legal system as enforced in British India rather own cultures were followed.

Indian Army and **British Army in India** also represented two different segments of the army. The regular British army regiments had a tenure of duty in India spanning over five to seven years, they had exclusive all British ranks.; they were known as British Army in India. The standard army which took its birth with the BEIC comprising of British Officers and native other ranks was known as Indian Army`. Native officers were given commission after First World War. Indian Army had its own commander in chief; there was some discrepancy among the allowances of both armies.

Durand Line marks the Chitral's Frontiers with Afghanistan in North and West

Durand Line & Chitral

In 1893 the agreement was signed by the Afghanistan and British India to have a demarcation of their respective international boundaries.

Chitral differs from other parts of Durand Line because the River Chitral after traversing the territory flows

10 NWFP, Imperial Gazetteer 1905, p-26.

Chitral Fort and Mehtar's Palace. Scene of 1895 siege

into the Afghanistan, there are 29 passes that leads to Chitral from Afghanistan staring from Darwaza Pass in the north to the Arandu or Aranwali in the south. Amir Abdur Rehman had agreed: "The province of Wakhan.. had come under my dominion... I renounced my claims on… Dir, Chilas, and Chitral to British for protection .[11]. Amir reluctantly took Wakhan but was bitter in giving up his claim on tribes and territories adjacent to River Kunar & River Kabul but in the end he put his signature on treaty which can be regarded as one of the most significant event of 18th Century. A masterpiece of diplomacy because British were weary of the history and thus lacked the conviction and force to impose their will upon Amir yet in the end Secretary of India's Foreign Affairs , Mortimer Durand was able to draw the line which ensured India's natural borders extending as far west as to have physical control of passes leading into India. Chitral physically was not affected by the Durand Line rather it was a blessing in disguise, it now physically separates them from Kabul's culture. Durand had kept the Arandu Pass as the boundary thereby cutting off traditional route of Kabul towards Kashmir.

Trouble initially started with the demarcation at Asmar{close to Arandu} within Afghanistan. Umra Khan of Jandol was another key player he was the son in law of Mehtar Aman ul Mulk, pathan by race he was the chief khan of Lowari Pass. After Aman's death the bloody intrigues for the Mehtar ship started among the various sons of late Aman ul Mulk {he had eighteen sons} finally it was the British weight which made Afzal Ul Mulk triumphant in 1894, it was short lived as Kabul backed Sher Afzal soon took the reins after bloody coup. British now backed Nizam Ul Mulk another son of late Aman ul Mulk, this resulted in Sher Afzal leaving the area as he had entered through Doeah pass back into Afghanistan.

A British political agent George Robertson was stationed at Chitral. He started his move in January 1893 towards the Chitral from Gilgit Agency with 50 men of 15th Sikh Battalion, Captain Young Husband a famous traveller was his assistant along with Lieutenant Gordon and Bruce. A native by the name Abdul Hakim also accompanied the Robertson. He arrived at Chitral and established himself, he was now the British representative and his main task was to act as a listening post on the northern gates of India.

Chitral Siege; 3rd March – 19th April, 1895.

After the death of Mehtar Aman Ul Mulk on 30th August 1892, anarchy erupted into Chitral. Prince Afzal ul Mulk , one among his eighteen sons took the reins of power after killing his three brothers and declared himself as the Mehtar, Prince Nizam ul Mulk the eldest son trekked to Gilgit to seek British help in getting Mehtarship. Meanwhile , Sher Afzal one of the brother of late Aman Ul Mulk attacked Chitral and killed Afzal ul Mulk and declared himself as the Mehtar. Amidst all this chaos British provided military aid to Prince Nizam ul Mulk who marched from Gilgit and defeated Sher Afzal in Battle of Drasan, Sher Afzal escaped to Afghanistan. British recognised the winner as the Mehtar of Chitral. Mehtar Nizam had little time to enjoy his labour, two years of rule, but in this brief period he invited British military mission to visit Chitral resultantly in May 1893 a small detachment with Captain Young husband was left at Chitral. On 1st January 1895 Mehtar Nizam ul Mulk was shot dead at Broze while hawking on the instigation of his nineteen years old half brother Amir ul Mulk[12]. Lieutenant Gordon the assistant political agent at Chitral was not certain about the impact and sought clarification from his superiors , in the mean time the young rebel lost patience and looked upon his sister's husband ' Napoleon

11 Ibid,pp,134-144. Also see,MuhammadQaiserJanjua. "In the Shadow of the Durand Line: Security, Stability, and the Future of Pakistan and Afghanistan". Naval Postgraduate School, Monterrey, California,pp 79-82.

12 Lieutenant Colonel Afzal Khan , Commandant Chitral Scouts 1971-1973. *Chitral & Kafiristan a personal study*{ Ferozsons, Lahore, 1975}, pp, 12-17.

of Pathans' Umra Khan of Jandol for help. Umra Khan a Pathan , had already descended down from Lowari along with his 1200 followers few days before the assassination of Mehtar Nizam ul Mulk; his declared aim was to eliminate the Kafirs of Kafiristan, he tried to bully the young Mehtar Amir ul Mulk in joining hands with him in such a noble cause but Amir refused resultantly Umra Khan laid siege to Drosh fort {also known as Chitarli Fort} which fell to him on 9th February 1895, Umra Khan also had the Chitral Fort in Narsat across Arandu inside Afghanistan .

Lieutenant Gordon in a counter move sent a SOS message to Major George Robertson the British political agent at Gilgit some 220 miles away. Major George marched from Gilgit with an escort of 400 troops, still believing that everything will be settled soon. George Robertson had negotiations with Amir Ul Mulk, who demanded immediate recognition of him as Mehtar for the cessation of hostilities; Umra Khan simply ignored the warning from George Robertson to leave the Chitral. The powder keg finally exploded on 28th February 1895 when Umra Khan Jandol and Prince Afzal Shah joined hands to physically evict the British. Mehtar Amir Ul Mulk was already placed under close arrest inside the fort by Major George Robertson and fourteen years old Shuja Ul Mulk was installed as Mehtar of Chitral. It was pure politics with a spice of religious fervour. It also speaks about the military culture of the area as in 1892 the Hunza & Nagar had fought a hard pitched campaign against the Colonel Algernon 's Gilgit Agency troops to preserve their integrity and suzerainty however in the end they went down fighting; now it was Chitral's turn. Kabul did not interfere openly in Chitral's affairs but it was covertly supporting Umra Khan and Kashmir State was fully backing the British.

The British officers[13] { Major Surgeon George Robertson, Captain Colin Campbell, Captain C. Townsand, Lieutenant H. Harley, Captain Surgeon Whitchurch, Captain Baird and Lieutenant Gordon} and their escort which comprised of 100 Sikh sowars of 14 Sikh Regiment and 300 sepoys of Kashmir infantry apart from 27 servants and 11 followers left Gilgit on 15th January 1895 and were inside the Chitral fort on 31st January along with 52 loyal Chitralis. By evening 28th February 1895 they were threatened and they timely got behind the safety of the fort before Umra Khan & Sher Afzal close the rope around their necks. Captain Campbell a hot blooded Scott overrode the aged wisdom of Major George Robertson and rode out of fort on 3rd March 1895, accompanied by Captain Baird his best friend with a sizeable force for an encounter with the enemy. By evening he was wounded along with thirty two other fellow soldiers and souls of twenty three loyal soldiers had already departed while fighting for survival on the battle field. On 4th March , Captain Baird took his last breath inside the fort, he had sustained serious wounds in yesterday's encounter 500 odd demoralised men were now compressed within the twenty five feet high and eight feet thick walls, measuring eighty square yards; it was a living cold hell. Food was short, heating arrangements barely sufficient, fodder insufficient but ammunition was enough, major flaw in the defence were the thick, high trees around fort which allows good sniping positions to enemy. There were loyal troops pouring in from Gilgit, 102 were at Mastuj and other hundred odd were enroute to Mastuj from Gilgit with another 100 in reserves at Gilgit.

Reshun & Kuragh . Chitral Fort held out for well over a month under persistent dangers waiting for relief. The first disaster came from unexpected corner, the follow up troops{60 soldiers, 40 coolies and two British officers} at Mastuj while marching towards Chitral on 7th March with 40,000 rounds of ammunition were ambushed at Reshun{near Booni}. For a week the two officers Lieutenant Edwardes & Fowler held out at Reshun, Lieutenant Fowler was already wounded on the very first day of ambush. In the end both officers were made hostage by Sher Afzal which further

13 Major Muhammad Nawaz {retired} *The Guardians of The Frontier , The Frontier Corps NWFP*, { Headquarters Frontier Corps, Peshawar, 1994}, pp 227-267. Also see *Chitral Story of a Minor Siege* by Major George Robertson.

Sitting: (1) M.J. Ghulam Dastagir (2) **Mehtar Shuja Ul Mulk** (3) Inayat Khan Lal (4) A.P.A Chitral (5) Un-Known
Standing: (1) Sultan Shah (2) Un-Known (3) Un-Known (4) Ataleq Surfaraz Shah (5) Hakim M. Sharif Khan (6) Mubarak Shah (7) Yaqoob Khan Lal

aggravated the situation for Major George Robertson. On hearing about the ambush at Reshun, the remaining British troops at Mastuj under Captain Ross and Lieutenant Jones also marched towards Reshun on 8th March with 93 soldiers and one native officer. Close to Reshun, Captain Ross left a party at Booni and went ahead with 33 soldiers to relieve his besieged brothers in arms at Reshun, they reached Kuragh which is short of Reshun , Captain Ross was also ambushed rather trapped in a narrow defile , he never made out of the trap alive despite heroic deeds by Lieutenant Jones. Captain Ross died on 10th March , only Lieutenant Jones and seventeen men survived the ordeal[14].

Lieutenant Colonel Kelly's March from Gilgit - Chitral, 1895

It was on 7th March 1895 that world came to know about the Chitral and it became an instant hit with newspapers who are always looking for such episodes. The relief forces came from Gilgit{Colonel Kelly} and Peshawar{ Major General Sir Robert Low with twelve infantry battalions, two cavalry regiments, four batteries of mountain artillery with usual supplement of Sappers & Miners}. It was almost replica of Kabul of 1842, only this time the residency was still intact and waiting for relief forces. It was this suspense which created the headlines in England and in Empire which made Chitral famous in military history. Everything was in favour of the besiegers especially the weather and the geography. This was the first British military expedition beyond Peshawar towards Chitral; there was no knowledge about the area . Resultantly on 21st March 1895, Lieutenant Colonel Kelly was ordered to move immediately from Gilgit 220 miles in east towards the Chitral and wait for the Peshawar Division. Colonel Kelly had four hundred men of Pioneer battalion who were primarily enrolled and trained as 'road constructors', Kelly had to cross the 12500 feet Shandur Pass almost 135 miles west of Gilgit in winter and Sir Low had to negotiate the 10500 feet Lowari Pass at the end of his 200 miles march crossing passes inhabited by hostile and militant tribes; a logistic nightmare. Kelly crossed snowline at Ghizar on 31stMarch and Shandur Pass on 4th April, a march worth appreciating. 13th April sunrise saw Kelly crossing Mastuj River, no serious resistance was encountered so far by him except the weather.

However at Nisa Gol which is eight miles down stream of Mastuj , a decisive encounter took place against rebel Isa Khan and his 1200 men on 14th April; Kelly was able to advance forward shrugging off Isa Khan. 17th April 1895 stands out as a day of envious bravery and chivalry , Sher Afzal had decided to physically attack the fort and for this a mine was being put under the fort wall which was detected in the nick of time about twenty meters away from the wall. The complete defence of the fort was in jeopardy, only a quick and brave action could save the inevitable. Lieutenant Harley led the raid to destroy the mine and in the process this young man sacrificed his own life to make sure that the mine is destroyed; he succeeded in his mission. Umra Khan fled from the scene as soon as his last trump card of mine was blown into air. Kelly was able to ford the River Chitral on 19th April 1895 thus ended the siege of Chitral. Both these columns in the end reach their destination the Chitral Fort, an impressive display of military leadership, planning, discipline, training and morale. Afzal Khan and Amir Ul Mulk were arrested and deported to India

In the end it was lack of leadership on part of Umra Khan and Mehtar along with real politicks which allowed the Kelly's force to reach Chitral in the nick of time to save the day; Low took some time to reach Chitral.

The classic act of frontier romance, a mountainous kingdom besieged by the wicked ruler and tribes and held out by a brave and loyal force of natives led by British officers, few casualties, no major clash or massive bloodbath. It was only at Chitral where after 1857 that the act

14 *Chitral & Kafiristan, a Personal Study*, p-42.

Hanging Wooden Bridge over River Chitral was constructed in 1899.

of Lucknow Residency was re-enacted with a happy ending, that is how Chitral Siege has gone down in the frontier history. It was different from Hunza & Nagar campaign as no Victoria Cross was awarded. It had a strategic implication had the Umra Khan of Jandol the most able and famous of the Pathan Khans been successful in his campaign to have Chitral under his influence then the British would have received a serious dent in the Forward Policy

In the aftermath of Chitral Siege of 1895, a new political agency was formed known as , Dir & Swat political agency with its headquarters at Malakand, Chitral was added to it a year later in 1896. Thus an assistant political officer used to manage the affairs of the Chitral; before this it was the political agency of Gilgit which was looking after the British interest in the state. It should be kept in mind that in 1891 under the Mehtar Aman ul Mulk the Chitral had agreed to hand over its foreign affairs mainly the relationship with the Afghanistan to the British.

Chitral Levies were raised in 1899 by the assistant political agent at Chitral; it was his personal escort and force to implement the orders, it had a strength of 100 men . Mehtar had his own bodyguards numbering over 200 and there was no regular army of Mehtar; there was one British army unit along with two guns which was stationed at Chitral. They had built a new fort a mile west and up of old fort which was besieged. Chitral Levies thus perform the duties of border military police as well, Ziarat, Mirkani and Arandu were the three initial posts of Chitral Levies.

Grave of Captain Baird who died on 4th March 1895 during siege of Chitral

Chapter Two

Chitral Scouts -1903- 1947

Chitral Scouts {CS} were raised in 1903 on the recommendation of the political agent Captain McMahan. At that time Chitral, Drosh, Chakdara, Malakand, Drazinda and Jandola in South Waziristan were the only places in agencies where there were British regular army battalions stationed. That speaks itself of the importance of the Chitral, Dir and Swat agency. The strength of British garrison at Chitral had been reduced to one infantry battalion along with sappers and a section of artillery. This strength remained at Chitral till 1942. Thus Chitral Scouts were raised to protect the line of communication from Lowari to Drosh; moreover in Upper Chitral it was to act as listening post.

Captain Grady the First Commandant of Chitral Scouts

The Chitral Scouts were raised as classical militia; comprised entirely of local population which would train only for one month a year and rest of the year perform their normal task. This is what Machiavelli[15] had propounded in 15th century and it was adopted in Chitral in entirety.

Strength was 1200, which means that this number will be given training, they were called for training and manning of post in batches of two hundred for a duration of one month each year. They were provided with Martin Snider rifles but were not allowed to take them home, ten rounds per rifle per scout was the first line ammunition. Two British officers remained in command. Headquarters were at Chitral where a fort was already constructed in 1895. Pay and allowances were shared by the political agent and the Mehtar Shuja Ul Mulk, who was also installed as the honorary commandant of Chitral Scouts, Shuja remained in this honorary position till 1936 when he died after a rule of 41 years. Thus his stint as honorary colonel commandant spans over 33 years the longest in Pakistan's military history. It was a continuity of history that traditionally the Mehtar was installed as the honorary colonel commandant. On raising, the initial batch of scouts mostly from Upper Chitral were given extensive training in drill for which the drill instructors from the regular British army unit stationed at Chitral were borrowed. Firing was the main thrill for the recruits if they can be called as recruits, air was informal and so was the routine. Polo was a major attraction in the evening when the scouts would show their prowess. Watching them playing polo was a treat and lesson in war itself, fearless gallops after a ball the control over the ponies and coordination between the hand and eye while galloping speaks volumes of military traits; it was only a question of harnessing these.

---

15 *Art of War*, Nicollai Machiavelli , translated from Latin.

CHITRAL SCOUTS
CHITRAL SHIKAR CLUB
RECORD OF BIG

Ski & Shikar Club Chitral Scouts 1905

Chitral Scouts at firing range 1906

Mehtar meanwhile initiated a programme of openness; he was invited and visited Calcutta in 1899, at Peshawar Vice Regal's Durbar in 1902 and at Coronation Durbar at Delhi in 1903. Mastuj and Upper Chitral including Laspur and Yarkhun were made part of Chitral territory with Mehtar of Chitral entering into a pact with the natives in 1909. In true sense the state of Chitral took birth after this. For centuries both halves have retained their identity, Mastuj although ruled by the same bloodline as in Chitral had political & military rivalry with the main town Chitral; presence of overwhelming Ismailis was another key factor. In the end the political agents of Gilgit and Chitral mutually agreed upon the interstate boundaries which roughly runs over the Shandur Pass. Gilgit had more complicated political system, it was part of Kashmir State yet the frontier fiefdoms of Yasin & Gupis were relatively independent in nature.

Chitral's Annual Relief.

British garrison which was stationed at the Chitral amounted to two infantry regiments, one company of miners and sappers, one section of two guns; deployed at Chitral and Drosh Killa . A line of communication was opened from Mardan through Malakand, Chakdara, Dir via Lowari to Drosh ; it was the responsibility of Khan of Dir to keep it open in his area for which he was given subsidy as well. Dir Levies were raised in 1897 for this purpose they were under command of the Khan of Dir. Similarly levies were also raised at Malakand known as Malakand Levies. This garrison was annually relieved utilizing this route. It was only in 1936 that the first ever relief using motor vehicles was utilized, the road from Chakdara was vastly improved by then still Lowari had to be crossed on foot. RAF from 1929 onwards started providing air lift and in 1937 an aerial relief was carried out in the winter when passes remained closed. Chitral Scouts initially had to look after the three posts namely Ziarat, Mirkani and Arandu.

Chitral Scouts 1910

Regiments stationed as far as Bangalore were moved to Chitral. Some regiments were required to adopt route march till a railhead and then to Nowshehra, where as the relieved battalions had to march from Drosh till Nowshehra. In most of the cases the relieved battalions were stationed at Abbottabad after their tenure at Chitral. The Peshawar Divison was responsible for the conduct with general officer commanding required to be physically present at Chakdara for the whole duration which took ten days both ways. Political agent of Swat, Malakand & Chitral was responsible for the local provisions of grain. Ice camps were also established on the advice of medical officers. Scale of ration and equipment was also altered; for troops going to Chitral were authorized to have three pair of shoes and two pair of socks. Importantly great coats were also issued individually otherwise at normal stations they were authorized 33 coats in one infantry company. Followers were also allowed to have 20 seers of luggage and they were issued one pair of socks apart from the No.3 Shoe.

Third Afghan War 1919

The war itself was not focused on Chitral sector, this sector differs from all other in a sense that it is inhabited by non Pathan tribes who do not share the common language with rest of combatants of the war. However purely from military geography Afghanistan had much more chances of success here in Chitral than anywhere else. Afghanistan had kept its words during the Great War 1914-1918 but then the chain of events inside Afghanistan necessitated launching of a war' Jihad' by Kabul ; which it did in the summer of 1919 with great dexterity and achieved stunning results in Waziristan sector. Chitral in the summer of 1919 was a peaceful garrison, the lone infantry regiment was stationed at Drosh, Chitral Scouts itself had only one company in the summer to train with or for manning of the posts. May has been a month of almost all major events in subcontinent thus it was in May 1919 that Amir Amanullah broke his relations with British India because it had delayed in accepting his kingship and more so politically to have the public support in his favour which is so vital in country like Afghanistan and what better way than to have Jihad; tribes became electric with the prospect of an all out war against the infidel British. Mehtar of Chitral also received one such firman from Amanullah on 8th May 1919 through the military governor of Afghan Asmar province Brigadier Muhammad Usman Khan; Mehtar Shuja Ul Mulk rejected the offer and kept his side of pact with the British intact. It was a very wise step of diplomacy. The war bugle was sounded and Chitral prepared itself for an attack by Afghanistan.

Chitral Garrison comprised of 1/11th Rajputtana Regiment with 450 bayonet strength it was commanded by Lieutenant Colonel F.C.S. Samborne who was also the commander of all armed forces in Chitral which comprised apart from his regiment, a company of sappers and section of guns ex 23 mountain battery, along with 1000 scouts of Chitral; they had been called up for the service. Chitral Scouts were being commanded by Captain Crimmin with Lieutenant Byres as the British officer.

Group Photo of Chitral Scouts Officers and Political Administration, taken at Drosh 1911

Chitral was divided into eight districts each under an 'aatlique' responsible for collection of revenues and also head of the local scouts, it was his responsibility to collect the men for training and send them to the Drosh . Under every aatlique there was a 'charwelo' responsible for a group of villages within a valley, each village itself had the Baramush the head . It was the Baramush who was the first ladder of the Chitral Scouts organization. Now these aataliques gathered the trained men . Chitral Scouts were under the political command of assistant political agent Major N.F. Reilly who started mobilizing on 5th May 1919.

Company of scouts was initially deployed at Galapach six miles downstream of Mirkani, aim was to keep an eye on the Arandu pass. Mehtar on the other hand had sent his various sons along with his body guards on various passes also. Initial Afghan movement started from 12th May onwards, they captured Arandu and soon the Scouts position at Galapach was over ran by 600 strong Afghans, scouts retreated to Mirkani. Afghans were certainly moving forward with an aim to capture the Mirkani and close the Lowari Pass.

Clash at Mirkani.

On 14th May 1919, Major Reilly along with two companies of Chitral Scouts arrived at Mirkani from Drosh. The retreating scouts were also harnessed and together these three companies put up a courageous attack on the Galapach position and reoccupied it; fighting continued the whole day. In another classical display of fighting the scouts along with 50 men of Mehtar bodyguard under command Major Reilly put up another attack on the Kauti feature which was across the river and source of permanent irritation for the Galapach position the objective was achieved with in hours on 16th May. Afghan strength at Arandu was estimated to be over 600 supported by four artillery guns and a large tribal lashkar, apart from Arandu the Afghans were in control of Bashgol and Birkot valleys

Chitral Scouts during Third Afghan War 1919

On 21st May 1919 in the overall plan the one company of Chitral Scouts were amalgamated with the Mehtar's bodyguard in an attack on the above mentioned positions. Two companies of Chitral Scouts under command Lieutenant Byres of Scouts pressed forward on the right bank and two companies on left bank under their commandant. Subsequently Byres moved forward , on 22nd May he had a temporary bridge thrown over the river in which the scouts were expert in doing so, Lieutenant Byres move ahead to face Afghans at Darashot but to his dismay they had vacated the post before his arrival. {Almost hundred years later in 2011, Afghans again attacked this post.}

Battle of Birkot - 23rd May 1919.

Almost entire British garrison at Chitral was present at the Battle of Birkot, a small town inside the Afghan province of Asmar where the bulk of Afghans were concentrated. Battle opened up on 23rd May at 0700 hours with own artillery opening up, the Rajputtana and

Chitral Scouts at Battle of Birkot 1919, captured Afghan Flags in inset

Captured Afghan Guns

Chitral Scouts carried out the advance and by 1400 hours the Afghans started retreating from Arandu. Afghan resistance was mainly in the form of sniping. Chitral Scouts lost eight men and another 23 were wounded. After the action the Afghans started reinforcing themselves thus the political administration of Chitral very wisely decided to retreat back into own areas thus the two companies of Chitral Scouts were left at Mirkani Piquet and remainder took position inside Drosh Fort along with other elements.

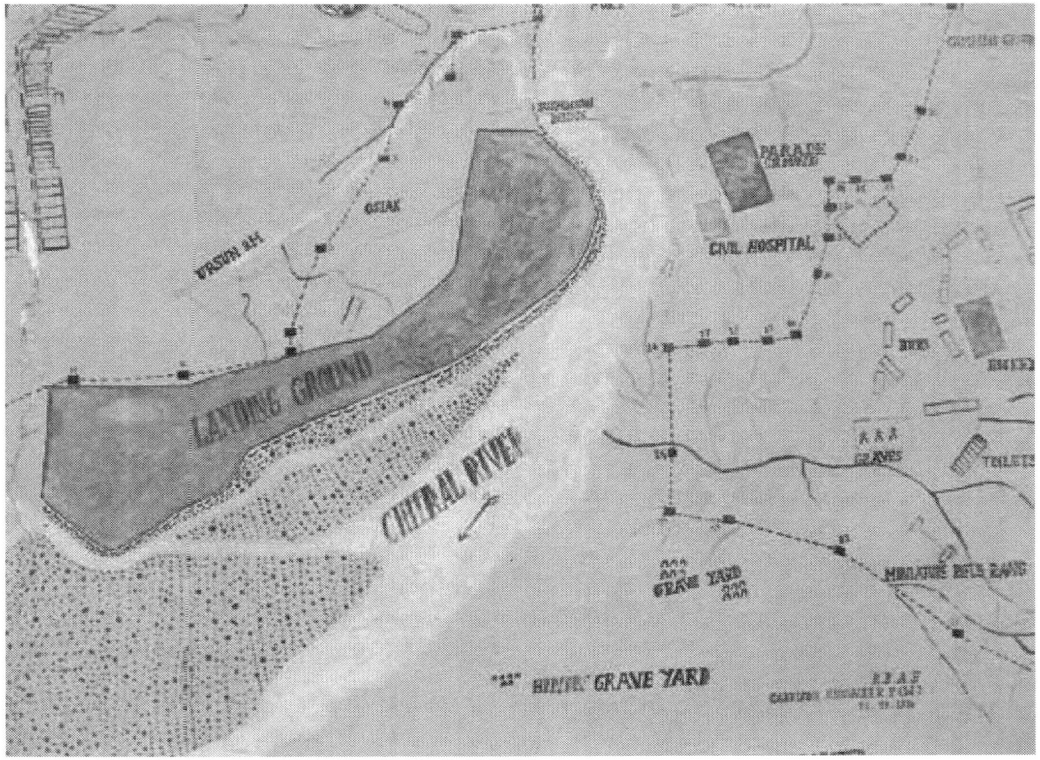

Landing ground Drosh 1930

On 3rd June 1919 the armistice was signed between the British India and Afghan government. However the situation at Chitral remained precarious with Afghan General Wakil Khan planning to move into the Chitral; thus two companies of Chitral Scouts and few men of Mehtar bodyguards were stationed at Ayun and one company was also sent to Lotkoh Valley. On 17th July 1919 an Afghan attack came upon Bambouret Valley via Zanor Pass and advanced to a mile of the Bambouret Village. Afghans had a complete control of all passes starting from Zidig Pass in the north to the Brambolu Pass in the south. Lowari Pass was also threatened by 500 odd Afghan tribesmen with menacing posture towards the Ziarat Post. Lieutenant Byres with two companies of Chitral Scouts physically attacked this conglomeration of tribesmen and pushed them back. The situation in Chitral started coming back to normalcy after signing of the treaty at Rawalpindi on 8th August 1919; thus ended the first blood of Chitral Scouts from which it emerged as a victor. Three scouts were awarded with posthumous award of Indian Order of Merit, one with Military Cross, one with distinguished Service Order and two each with Indian Distinguished Service Medal and title of Khan Sahib.

More importantly the concept of Scouts in Chitral, so different from the other militias proved a success. Unlike North Waziristan Militia and South Waziristan Militia along with Khyber Rifles where mass defection took place, there was not even one defection among the ranks of Chitral Scouts. Being entirely composed of one qaum it had its own advantages which proved so good in this conflict.

Chitral Scouts - 1930

The Mehtar was knighted after the Third Anglo-Afghan War , his subsidy was also increased and above all he was entitled to 11 guns salute now. British thus in an effort to lessen the financial burden upon the state undertook the proposal to reorganize the Chitral Scouts in this background. There were 9 companies of Chitral Scouts at the end of the 1919 but these were gradually reduced and by 1930 there were six companies which were further reduced to 4 companies. The reduction was carried out after a detailed study[16]. Each company had four platoons with two subedars, eight havildars, eight naiks and 200 temporary naiks or soldiers. Thus each subedar to command two platoons or half company with a havildar designated as platoon commander. The subedar and havildar were bound to undergo one additional month of training apart from the one month which they were supposed to undertake with their company. Each company itself to under go one month training. On the financial side a reduction of 109 men all ranks was carried out. Companies were named as Lasper Mulikho company, Turikho company, Mastuj company, Kuff Company and Lutkoh half company.

Chitral Scouts had strength of four and half companies at the beginning of 1931. Pay in 1930 was as follow, a subedar was getting rupees 50/ per month where as a havildar was getting rupees15/, a naik rupees 12/ and sepoy rupees 10/per month. Ration allowance was given at a rate of Rs 9/ pm, the subedar major apart from his pay was given an annual allowance of Rs 25/. Clerk was a highly paid trade with a pay of Rs 120 per month, armourer was paid Rs 45, religious teacher was getting Rs 10, and gardener was being paid Rs 9 per month. The strength of Chitral Scouts was as follow, eight subedars, 32 havildars, 32 naiks, 64 lance naiks and 736 sepoys. There were four chowkidars as well who were being paid at Rs 15 per month, there was a pupil teacher also.

Gilgit Scouts, Great Game & Gilgit Agency. 1936

The political and military rivalry of European continent also had its effects on the sub continental politics, initially British had apprehensions about the Napoleon's advance towards the India which were later replaced by the Russians after the treaty of Vienna in 1815 A.D. By this time in history the area west of present day Afghanistan was a conglomeration of independent small valley states all Islamic in nature like Khiva and Merv[17], these were captured by the Russians therefore British in order to keep an eye on the affairs wanted to have listening posts thus the Gilgit campaign of 1892 followed by Chitral in 1895.

Lease of Gilgit, 26th March, 1935

On above mentioned date Maharaja of Kashmir Hari Singh leased the 'Wazarat of Gilgit' to British for sixty years with a condition that flag of Kashmir State will be flown at all the official buildings including the British political agent headquarters. Gilgit had a population of 78,314 souls in 1911 which rose to 89,697 in 1921 . Gilgit town and adjoining areas had 31,902, Hunza 13,241, Punial 6,108, Chilas 13,534, Nagar 13,672, Yasin 8,083, Kuh 2,808, Ghizar 4,112 and Ishkoman 2,986 souls in 1931 and collectively they were known as Gilgit Wazarat, the area given under lease thus

16 Report of Captain McCardy 1930, Chitral Scouts Archive

17 Anila, Bali. *The Russo-Afghan Boundary Demarcation 1884-95: Britain and the Russian Threat to the Security of India,* (School of Humanities, University of Ulster. Thesis submitted for the degree of D.Phil.1985). pp. 25-27. *Also see* J.A. MacGahan *Campaigning on The Oxus and the Fall of Khiva* (Simpson, London, 1874) pp 163 – 193.

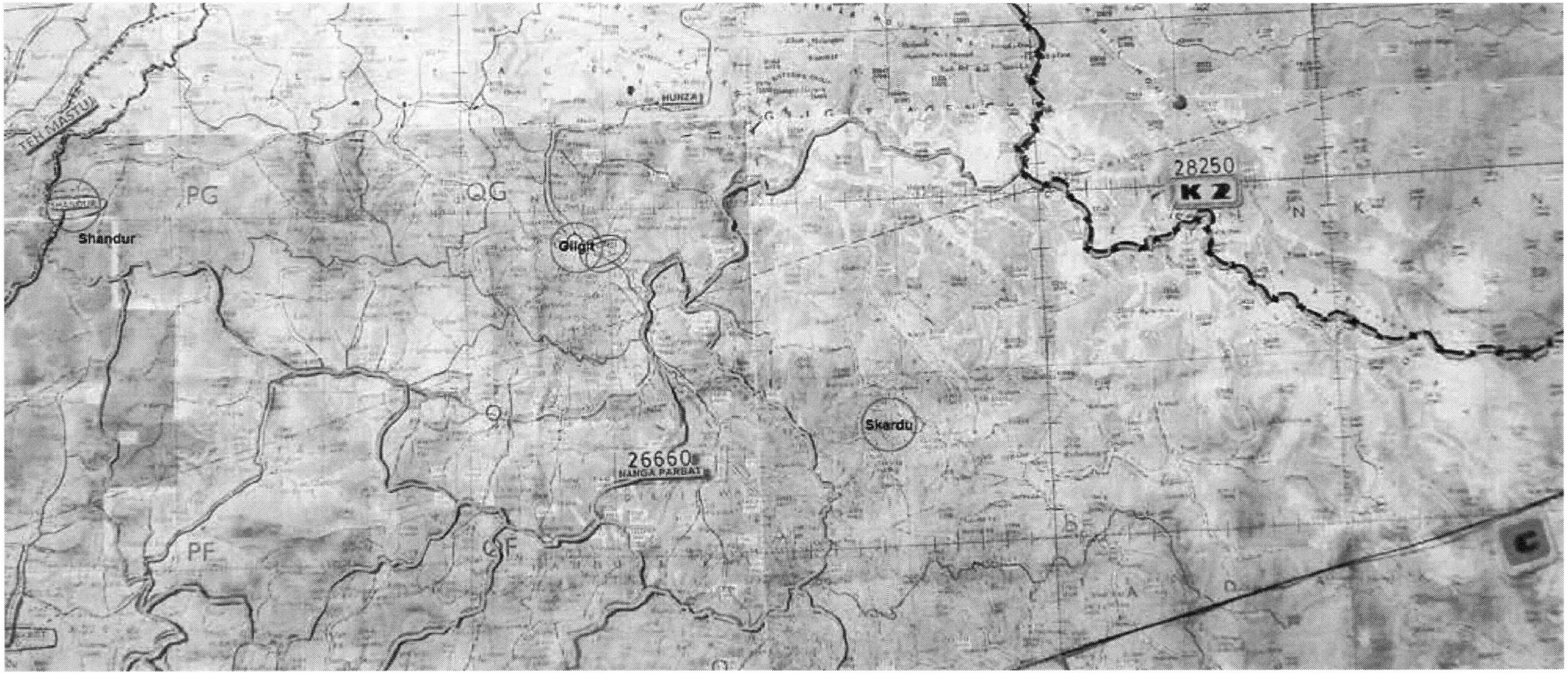

Million scale map of Gilgit and Baltistan

was known as Gilgit Agency, it practically amounted to all area west of River Indus[18].' Gilgit Agency comprised of Gilgit Wazarat made up of tehsils of Astor and Gilgit and Niabat of Bunji, Punial jagir, states of Hunza and Nagar, governorships of Yasin, Koh, Ghizar and Ishkaman and the republican communities of Chilas.[19]

Culturally the people of Gilgit are a martial race and shun manual work, they think very high of themselves and their ancestors and considers themselves as a race of rulers, they are brave and one cannot but admire their determination, their hair is usually black and sometimes brown, their

18 Military Report and Gazetteer of the Gilgit Agency and independent territories of Dariel and Tagir, {Government Monotype Press, Simla, 109}p-3.
19 Frontier and Overseas Expeditions from India, volume I, p-6.

eyes are either brown or hazel, voice and manners of speech appears to be harsh. Dard language with its three dialects is generally spoken through out Gilgit. Gilgit as a class are divided among five classes, Ronu the highest class followed by Shin{they abhor cow and as such do not drink her milk neither they burn her dung},Yashkin, Dooms and Kamins.[20]

Gilgit Scouts were raised in 1915, consisting of local natives {six hundred in number} commanded by the British officers and mainly funded by the J&K State; posting of British officers in Gilgit scouts was the domain of British Army and Frontier Corps had nothing to do with it. However now in 1936, under the lease the Gilgit Scouts came under administrative control of Frontier Corps and officers were inter posted among the corps. Gilgit Scouts similar to Chitral Scouts are non Pashtu speaking and have identical cultures with Chitral., there was no native ruling elite. Gilgit comprised of many small mountainous fiefdoms , thinly populated, economically poor. Its geography is also identical with Chitral, few miles west of Gilgit Airport the river confluence in same pattern as in Chitral. The upper Gilgit is also Ismaili followers by faith. In 1907 the men declared fit to bear arms were 5804 with 1500 in Hunza and 2000 in Nagar, 800 in Chilas and 1,000 in Yasin.

Gilgit was invaded by Gohar Aman of Yasin in 1840 , killed its ruler Shah Sikander and then ruled till 1842, when he was defeated in a battle by the Syed Nathe Shah the commander of Sheikh Mohi ud Din the governor of Kashmir under the Sikhs, resultantly daughters of rajas of Hunza, Nagar and Gohar Aman were given in marriage to Syed Nathe Shah. Karim Khan was made governor of Gilgit but soon Gohar Aman again revolted and in a famous battle cut Bhoop Singh forces into halves. In 1860 Ranbir Singh Maharaja of Kashmir sent an army under Devi Singh who captured Gilgit without any difficulty in 1860. It was devi Singh who carved a buffer state of Punial between Gilgit and Yasin. Soon Yasin was also invaded and Malik Aman son of Gohar Aman fled to Chitral. In 1866 Hunza was attacked by Dogras but failed to achieve the goals. In 1859 the very first British officer was posted at Leh to act as a listening post and from 1868 the British officers were posted at Gilgit with intervals for same purpose.

Darel, Tangir, Thor, Jalkot, and Seo were small republics on the right bank of the River Indus, below Punial. In 1866 Darel started paying tribute to Kashmir Durbar annually in terms of gold dust, collectively these are known as Shinkai Republics and also as Dardestan.These republics were dependent upon Punial and Yasin for grazing in summer. In 1895 Raja Pukthun Wali settled in Tangir after the turmoil in Chitral, he is regarded as father of Darel & Tangir Kingdom, recognised by British on condition of allowing them to built road in his area. Raja Wali was murdered in 1917.[21] Jalkot stands out as the bastion of native resistance to the Maharajja and British, from 1902 onwards their raids on Dogra posts increased starting with Kargah, a blockade on tribe was imposed which was repeated in 1924 also. Telegraphic communication with Gilgit were established in 1892,where as postal system was established in 1894. Relationship between Chitral, Gilgit and Kashmir remained turbulent. An extract from a communication by Munphool Pandit to the political department, India Office dating back to 1867[22] highlights that Gilgit did not had more than 1000 houses, length of valley from nort-south not more than hundred miles with a mean breadth of 20 miles, Yasin on west, Hunza and Nagri in the north & northeast, Darel in south and southeast and Balti in east

20 Munphool Pundit, relationship between Gilgit, Chitral and Kashmir, *Journal of the Ethnological Society of London{1869-70}. Volume 2, No.I*{1870}, pp.35-39. http://www.jstor.org/stable/3014434, accessed on 24th January 2012.

20 Gilgit The Northern Gates of India,, {Delhi, 1977}. P-18-19 also see Aithcheson, Treaties, Engagements and Sanads, p-12.

21 Gilgit The Northern Gates of India, pp.18-19.

In 1936 Nasir ul Mulk was installed as the Mehtar after the death of his father Shuja-ul-Mulk, the Kashmir Durbar was not informed about this succession , on which the Maharaja wrote a letter expressing his displeasure, soon the Chitral annual tribute paid to Kashmir under the 1914 Mastuj Agreement was also stopped; this tribute was never paid from 1892 onwards. Mehtar Nasir ul Mulk visited Kashmir in 1939 for a new agreement.

Chitral retained its strategic position, a road was greatly improved between the Chakdara and motor transport was used for annual relief till Dir. Commander in Chief paid a visit to the Chitral and Chitral Scouts in 1936. The road move was a tedious and dangerous manoeuvre with constant air support and piqueting of the route. Dir Levies were up to the task, air relief was carried in the winter of 1936-37 when aircraft taking off from Risalpur would land at Drosh within an hour. Heavy pieces of artillery were flown from Rawalpindi which included 3.7inch howitzer. For seven years Mehtar Nasir ul Mulk reigned Chitral and in his tenure the Chitral opened up a little bit towards the Afghanistan in terms of timber trade. Mehtar was an enlighten ruler having served with a British regiment for a year, he was the ideal; he politically united Chitral with Dir by marrying the daughter of Nawab of Dir. Mehtar Nasir ul Mulk was also given the honorary rank of major in the British Indian Army.

Hall name after Sherpa Tenzing Norgay

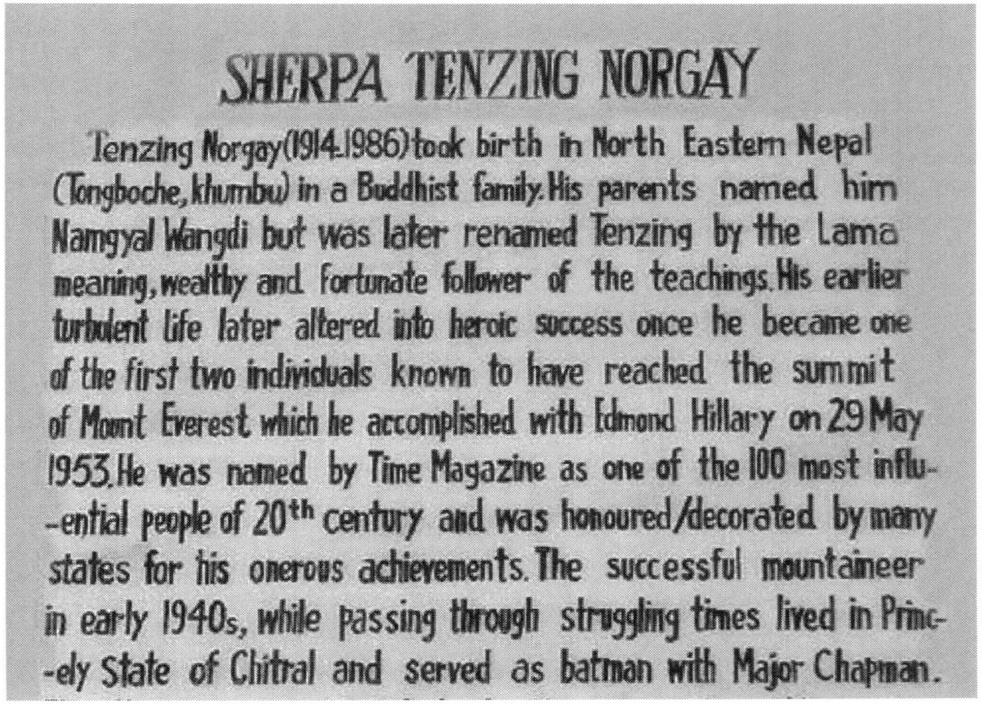

Chapter Three

Chitral State Scouts - 1942- 1956

In 1942 the last of British soldier left the Chitral, for they were required at more important places due to swinging fortunes of Second Great War which at time were placed heavily against the British. In the same year the nomenclature of Chitral Scouts was changed into Chitral State Scouts, it now became the regular forces of the Chitral State and placed under the Frontier Corps administrative control. The two northern most gates of India{ Chitral & Gilgit} thus became the responsibility and domain of Frontier Corps.

The strength of Chitral State Scouts was increased to 10 companies with an addition of one section of 2.75 inch Vickers machine gun. The strength was first increased from four companies to six in 1935. More importantly the scouts now came under the control of Frontier Corps for administrative purposes. The political agent of Dir, Swat & Chitral Agency still retained their control but now the posting of officers became a FC affair rather than a military. The first fruit of this was the training of Chitral State Scouts troops on the machine gun which was provided by the instructors from Tochi Scouts & South Waziristan Scouts ; moreover native officers of Chitral State Scouts had to spend three months either at Tochi or at SWS to learn the fine tricks of frontier warfare . Overall the Chitral State Scouts still retained their individual flair from the other corps of Frontier Corps; they maintained their one qaum recruitment and also the unique system of mobilization and one month training.

Pakistan - 1947

The political history of Gilgit and Chitral is almost interwoven and so is the military history of Chitral State Scouts and Gilgit Scouts. In 1947 when the question of accession of states to Pakistan arose, there was a feeling among the ruling elite of Chitral to be interlink with Afghanistan more than with Pakistan basing upon the geography but the 3rd June 1947 plan had left no option to the states but to choose either between the India or Pakistan. Mehtar Muzafur ul Mulk had good relations with Quaid e Azam and Chitral acceded to Pakistan in August 1947; Gilgit has a different story.

Lord Mountbatten the last Viceroy of India had announced the plan on 3rd June 1947[23], the independent states that included Chitral, Hunza, Nagir, Kashmir and Jammu State had the option of joining either dominion or remain independent, however this last option was never seriously considered and all princely states were expected to join India or Pakisatn. There were no conditions on the ruler regarding his choice of accession, it

23 *Partition Of Punjab Official Papers* ,pp 45-65.

was presumed that logic would prevail upon all and rulers will follow the rule of majority. Kashmir and Hyderabad Deccan defied this logic. In June 1947 British abruptly terminated the 60 years lease of Gilgit from Kashmir , Gilgit thus again became Kashmir Darbar territory and Gilgit Scouts reverted back from Frontier Corps to Kashmir military authorities.

Operation Datta Khel - November 1947

On 3rdt November 1947 at Gilgit the 600 strong Gilgit Scouts along with their commandant Major Brown a British officer revolted against the Kashmir government[24] after it had signed a Letter of Accession to Indian Dominion rather than opting for Pakistan (as the lease had expired after the transfer of power, Gilgit and Laddakh including Skardu were now again part of the Kashmir). This is probably the only incident of any British officer in one hundred and fifty years in India to become a rebel against his very own superiors, the military forces of both dominions was placed under Field Marshall Auckinleck and all commanders of Pakistan and Indian armed forces were British.

Gilgit Scouts were raised as Gilgit Levies in 1889 and they took part in Hunza-Nagar expedition of 1891 under Colonel Algernon Durand, later in years they were part of Chitral relief force as well in 1895. In 1913 the levies were disbanded and Corps of Gilgit Scouts were raised. Eight companies of Gilgit Scouts were deployed along Chinese frontier during WWI, they also reinforced the Chitral Scouts in 1919 War with Afghanistan.

Chitral Scouts had this unique honour that the very operation which changed the geo-strategic position of Kashmir for ever was carried out by an officer who had served in Chitral Scouts before moving to join Gilgit Scouts in July 1947. The operation itself was named after Tochi Scouts post at end of Tochi Valley ' Datta Khel[25]'. Captain William Brown was commissioned in a Frontier Force Regiment in 1941, transferred to SWS , served in Gilgit Scouts , came back to Tochi Scouts and then to Chitral Scouts and finally took the command of Gilgit Scouts after resigning his King's Commission. He planned the operation on the instigation of Lieutenant Colonel Bacon who remained as Political Agent in NWA and was the Inspecting Officer of Frontier Corps in 1947. So far there is no empirical evidence to link Major Brown's act with army headquarters; however there is enough logic to convince that accession of Gilgit with Pakistan was conceived and executed by the Frontier Corps. Most notably by the Lieutenant Colonel Bacon with strong support from governor of province; His Excellency Cunningham. It is no secret that British had no soft corner for the Indian Congress or its leadership. To the Frontier Corps the stakes were high; they could not simply leave the events to take their natural course of action which Mr Atlee at 10 Downing Street was wishing.

Gilgit Scouts thus became the nucleus around which the Gilgitis, Baltis, Chilasis {all Muslims} attacked Sikh garrisons at Bunji and Skardu capturing Kargil there by cutting the lone Himalayan communication link between Kashmir and Laddakh. This war was independent of the resistance that was being waged all along the Kashmir's border with Pakistan.

---

24 Alastair,Lamb, *Kashmir Disputed Legacy* ,p,161

25 Major William Brown, *The Gilgit Rebellion*, {Ibex, Rawalpindi, 1998] p- 117. Also see Victoria Schofield *Kashmir in Conflict: India, Pakistan & the unending War* {London, Taurus, 2010} p-63.

Pakistan government sent Sardar Muhammad Alam and Major Muhammad Aslam as political agent and military commanders respectively. By the end of November 1947, Major Muhammad Aslam under the nume de plume of Colonel Pasha organized the available forces into four wings and decided to continue fight for freedom of Baltistan despite approaching winters. Colonel Pasha identified two axis along which enemy could send reinforcements- one was along Bandipura through Tragbal Pass, Minimerg and Gilgit whereas the other was along Zoji La, Kargil, Hamzigund and Skardu. The main task, however, was to liberate Skardu from the Dogra forces. Basing on this appreciation Colonel Pasha organized his four wings into two forces i.e Tiger Force and Ibex Force.

Tiger Force. It was commanded by Captain (later Lieutenant Colonel) Hassan Khan. Its task was to capture Tragbal Pass and check enemy reinforcement towards Gilgit. The force left Bunji on 3 March 1948 and by mid of March it was able to capture Tragbal Pass. In May 1948 enemy launched an attack with two brigades. The Tiger Force defended their position relentlessly but ill equipped and ill fed 300 troops could not hold on for long and had to withdraw from the pass on 16th May 1948. They, however, took up positions about 20 miles behind where they stayed on till cease fire.

Ibex Force. This force was commanded by Major Ehsan Ali of Hunza. Its task was to free Baltistan and advance right up to Kargil, Dras and Zoji La Pass to stop enemy reinforcement along Indus Valley. It left Bunji along Indus River on 29th January 1948. After clearing enemy resistance at Rondu the force encamped at Sondus in the foot of Kharpocha fort. Muslim elements in the Dogra force had silently joined Major Ihsan Ali and it was decided to attack Skardu on night 22nd-23rd February 1948. Everything went as planned but unfortunately the sentry who had to fire green signal fired red one by mistake and hence chance for a surprise attack was lost. Fierce fight ensued but Major Ihsan Ali had to withdraw to reorganize and regroup.

Enemy started reinforcing Skardu and a brigade size force was approaching from Indus Valley side. This force was successfully ambushed at ThorgoPari and huge cache of weapon, ammunition and ration was captured. Siege of Skardu continued till August 1948 since enemy had been supplying the garrison by air. On 14th August 1948, exactly one year after Pakistan's Independence, Skardu was surrendered by Dogra forces.

Eskimo Force. After unsuccessful attack on Skardu in February 1948, Colonel Pasha organized another force at Chilam Chauki under Lieutenant Colonel Shah Khan. The force was named Eskimo force. It had to capture Kargil, Dras and Zoji La Passes which was initially the task of Ibex force. This force successfully captured Kargil and completed the siege of Dras when they were joined by Ibex Force. These two forces combined played havoc with enemy and were able to capture Kargil, Dras, Zoji la Pass and had invested Leh as well. At this point in time, General Headquarters replaced Colonel Pasha with Colonel Jilani who decided to strengthen his positions at Kargil, Dras and Zojila and abandoned the idea to attack Leh. The major considerations for this were lack of resources and over dispersion of forces.[26]

Chitral State Scouts & Chitral State Bodyguards-1947

Mehtar of Chitral, Muzzafar ul Mulk declared jihad for the liberation of Kashmir, he had already signed an instrument of accession to Pakistan. The force structure at Chitral comprised of Chitral State Scouts and Mehtar's bodyguard. Chitral State Bodyguards were the personal and loyal troops

26 Shuja Nawaz, The First Kashmir War Revisited, *India Review, Volume 7, No.2, April-June 2008*,pp.115-154. Taylor & Francis Group.

of Mehtar, they were important and vital in a state where majority of rulers have been replaced with assassination. They were raised in 1909 and issued with rifles but ammunition was provided next year. The situation at Gilgit was not very clear because after first week of November 1947 the Gilgit Scouts had declared an independence which was very complex in nature and it died down after a fortnight. At this stage the Dogras and Indian Army was moving into the Kashmir. Thus the confrontation between the Pakistan and Indian troops remained confined to the east of Indus where as the area west of Indus were left to the Dogra and the Scouts to fight it out with winner taking it all. Situation in Guraiz (Astore Sector) and Skardu valleys was tense and precarious in nature, Gilgit Scouts under command Colonel Hassan Gilgiti were retreating. The enemy had made considerable advance across Burzil Pass. It is in this back ground that Chitral State Scouts and Chitral State Bodyguards entered into the arena. Chitral Scouts relieved Gilgitis in Kamri and Domel Sectors where as Bodyguards headed towards Skardu.[27]

His Highness, Colonel Muttah ul Mulk, Victor of Skardu

Ruler of the state and honorary colonel of Chitral State Scouts Mehtar Muzzafar ul Mulk sent his bodyguards to fight along with the Chitral Scouts. He sent two companies of bodyguards under command his younger brother Muttah ul Mulk; who was given the rank of colonel, other two brothers were also sent there. The first wave of mujahedeen from Chitral had gone in early days of October 1947 towards the Kashmir , this was comprised of lashkar from Drosh led by Haider ul Mulk, Amin ul Mulk & Saif Ullah Jan , out of which Saif embraced shahadat at Noushehra; these mujahedeen later came back to Drosh.

Chitral State Scouts two companies were also under the overall command of Prince Muttah ul Mulk with Captain Agha Asghar Ali{he was wounded and reverted back to Chitral after a month}, honorary lieutenant Abdul Rauf Khan and Subedar Jan Badshah. The Scouts companies were deployed and employed on the Guraiz-Kargil sector.

Gilgit after the freedom act of Gilgit Scouts was still under the threat of the Dogra forces stationed at Skardu and Leh. There was a company strength at Skardu under Captain Parbat Singh as a reinforcement to the Dogra Battalion, another Dogra officer Captain Kishen Singh was holding the Tsari Pass the mouth leading to the Skardu along the Indus going upstream. Brigadier Faqir Singh was the overall commander with headquarters at Kargil. The bodyguards led by prince himself a WW2 veteran of Burma, where he became a POW at the hands of Japanese, marched towards the Gilgit.

Prince Muttah ul Mulk was a classical princely officer {he married a Chinese woman while in Japanese prison during WW2}, brave and leader of men; he will always be remembered as the Victor of Skardu.

He reached Skardu via Deosai plains{15000 feet}, adopting and circumnavigating the Indian held Tsari Pass and thus descended down upon the Skardu City from north east where as the Dogras had their troops stretched on south east; the only track coming into the Skardu Bowl along the Indus upstream., he had achieved the surprise with his two companies of bodyguards and laid the siege, the Skardu garrison was held strongly by the

27 Yaqoob Khan Bangash, 'Three Forgotten Acessions, Gilgit,Hunza and Nagar', *The Journal of Imperial and Commonwealth History*, Volume38,No.1, March 2010,pp117-143.

Skardu town in the Indus River bowl with Deosai Plains 15000 feet, in the north. Chitral State Scouts & Bodyguards reached Skardu Fort {inset} from this unexpected approach.

Impregnable Kharpachu Fort perched and nestled on steep, rocky mountain overlooking the River Indus and River shigar on east and the town itself is spanned at the base of rock, offering the lone climb up.

Dogra army at the almost impregnable fort on Indus. There was nothing which could be done by bodyguards except to bleed the Dogra which they did by remaining steadfast and withstanding the few enemy air attacks on them. Every night these Chitralis would climb the rugged and difficult mountain terrain and try to reach near the fort. Mutath ul Mulk had sent a letter on 17th June 1948 to the besieged Dogra force at Skardu ' I advice you to surrender .. I take the responsibility of your safety'.[28] On 19th June the besieged commander Lieutenant Colonel Thapha sent his emissary, Sepoy Amarnath with a white flag to Muttah ul Mulk accepting surrender terms under Geneva Convention[29].

However 180 of his men refused to do so and tried to escape out of which 80 were captured and remaining died in the escape. Captain Ganga Singh was handed over to the natives of Skardu who executed him for his crimes of rape and turning the mosque into a brothel house. Chitral Bodyguards who stood out in the campaign includes Lieutenant Sardar Aman Shah, Kahdim Dastigir, Ali Dyar, Atai Karim, Ayub Khan, Abu Ilas Khan, Abdul Jihan Khan, Sher Arab Khan, Wilayat Khan, Mehrab Hussain , Muhababt Khan and Muhamamd Gul.There were ten shaheeds from bodyguard including Sirajud Din, Jeidi Kahn, Akbar Kahn, Sobi Khan, Dawa Paanh, Nawab Kahn, Barzangi Khan and Mirza Nadir[30].

Bodyguards were a pack of Chitrali commoners who were having insufficient clothing and food, *' one old shirt, torn shalwar, an vintage chooga{coat}....a shredded scarf which tried to cover their neck, feet from the customary socks, a torn chappali or old shoe which was kept working with innumerable laces or clothes wrapped around it'*.[31]

Colonel Mutath ul Mulk did not have a peaceful end, he was arrested on 12th September 1948 at Peshawar, tried and put behind the bars at Peshawar, he was later released in October 1949 and reinstated as Governor of Mastuj. Reason for his arrest was political in nature as he became too popular among the people of Chitral; rather than anything to do with war crimes.

Asmat Wali's Diary. Nursing assistant of Chitral State Scouts, Baba Asmat Wali died in 1988, he took part in Kashmir Jehad of 1948, more importantly he left a diary of the account. Distance from Chitral to Kamri was covered in 38 days, Asmat on 20th December 1948 notes in his diary ' saw Pakistan currency at a shop , Government of Pakistan was written on the currency note'. He also notes down the change of command at Gilgit Scouts where Lieutenant Colonel Abdullah Jan was transferred and Major Tufail took over the command of Gilgit Scouts. Asmat also mentions the Indian counterpart Colonel Suba Lal Chand who according to him was a shrewd man. But the most touching part of his diary is the account of an old woman in Minimerg, 'a rich widow who sacrificed over twenty five goats in four months for the scouts because they were extremely short on ration'. Asmat when arrived back as victor to his native town in November 1949 came to know that he had lost his sister in his absence she had died almost immediately when he had left for Jehad but mother intentionally did not let pass the information to him.

Naib Subedar Islam Shah, had joined the Chitral Scouts in 1932 and he fought extremely well in the Kashmir War for which he was promoted to present rank something which was purely reserved for the Kator family, on his promotion many other subedars of Kator family resigned. Khuda

28 Muhammad Yousaf , *Tareeq Baltisatn*, p- 268

29 Lieutenant Colonel Mirza Hasasn, *Shamsher say Zanjeer Tak*, urdu, autobiography, 1979, pp, 250-279.

30 Naik Iqrara Ud din Khusro, *Chitral aur Chitral Scouts Tareekh kaay ainey mein*, {urdu} unpublished, the writer is a naik in Chitral Scouts,

31 *Shamsher say zanjeer*, p-271.

Baksh Utrai was another scout who was promoted to the rank of Jemadar due to his performance in the action, he was the very first havildar major of the Chitral State Scouts

An interview with a veteran of First Kashmir War

Interview was conducted at Sonoghar in Mastuj Valley, Chitral. September 20th 2013.

'We set off for the veteran Nadir who was still sitting under the tree but there were two additional chairs also put there, I occupied one and Amin who also acted as the interpreter did other. The children all sat around, there was only the spring water flowing melody, the walnut tree had a big hole in it, there was sun and there

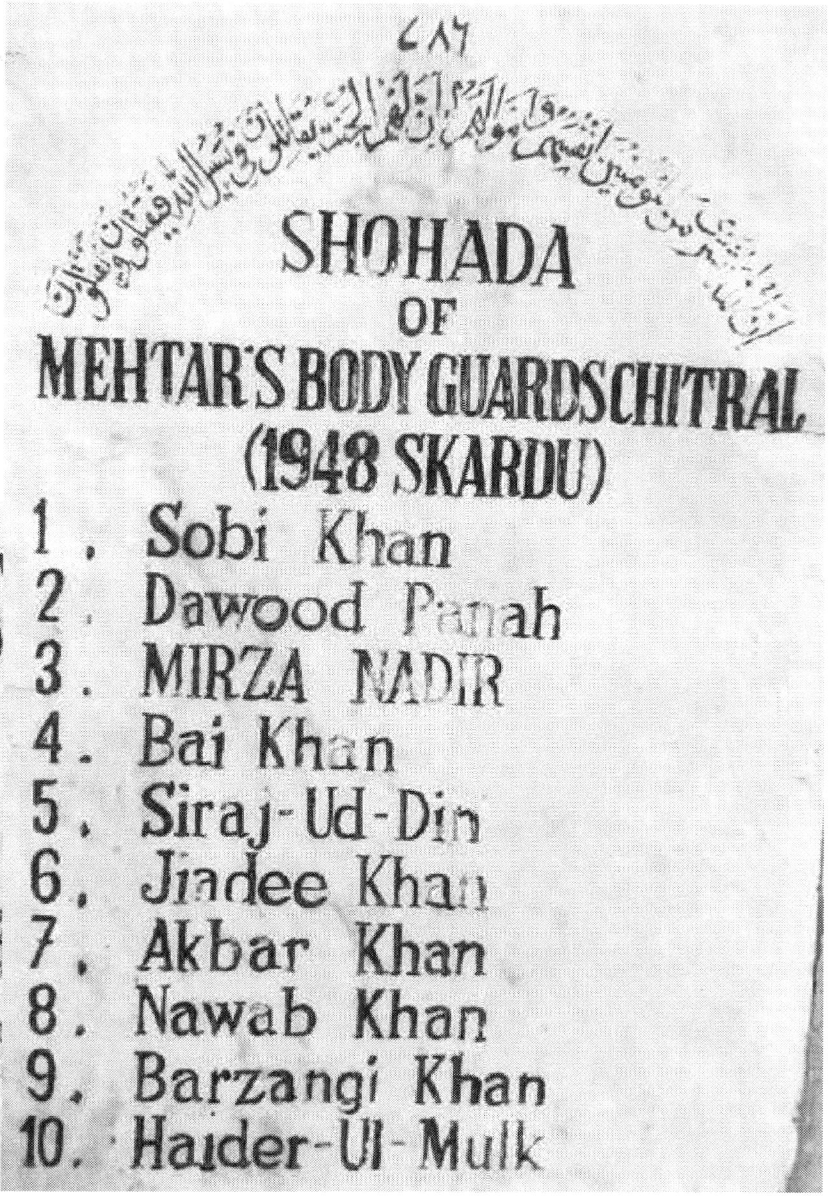

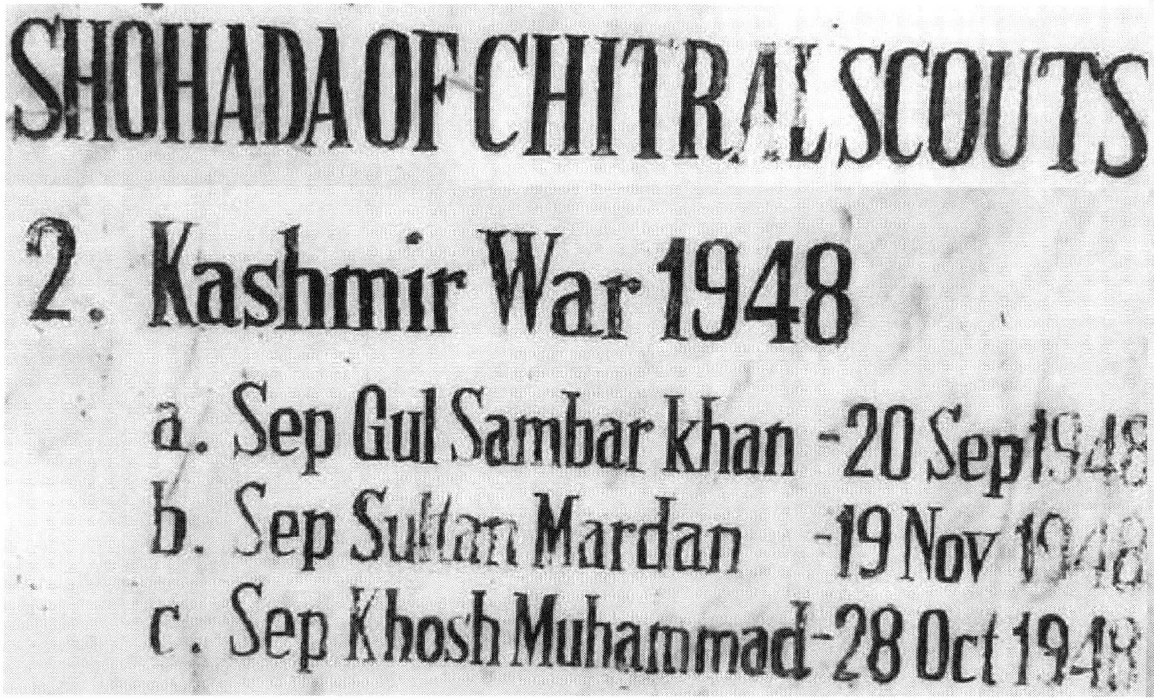

was shade also, in the close by distance were few fields having the sunrays falling in the middle and brightening up the already cheerful environs. Few women were standing at rather far distance out of ear shot looking after the grazing cows . Bit of silence as I absorbed the beauty and in any case you do not just go and ask the veteran a question, in most of cases it takes bit of time to understand each other especially for me to start the question, in this case when I asked him, when were you enrolled? And got a puzzled look, it became obvious that he does not understand Urdu thus Amin asked the same and that is how I got it. I could understand few words and thus make out the whole purpose, old man like so many other natives do not use hand signs frequently. Nadir was enrolled in 1945 in Chitral State Scouts at Drosh and in 1961 got his retirement papers from Drosh too.

During 1947-48 Kashmir War he was part of artillery battery, it took nine mules to carry one howitzer. They moved through the Shandur Pass towards the Gilgit amidst the snow and extreme freezing temperature; they had very little warm clothing with them as it was promised to be supplied at Gilgit. Nadir and the Chitral Scouts remain committed for another seven months. Nadir highlighted that the Chitral State Scouts and Chitral State Bodyguards fought the war separately. Nadir was deployed at Gurais with his guns for another seven months before they were pushed back. Nadir used to get rupees two only per month, mainly for purchase of milk, although the pay then was rupees 22/ but Nadir like all other men got it after coming home, in case of Nadir it was rupees 400/ quite a sum in those days. There was little ration and insufficient warm clothing at the front. Captain Rauf was in charge of the gunners, Major Mohiuddin who was the brother of Mastuj governor was in charge of the scouts contingent, thus in a way the princely family of Chitral were all engaged actively in the liberation of Kashmir. For ration Nadir went out on hunting and hunted markhoor, they ate the meat, preserved it, presented the head to the officer in charge and made shoes and coat out of the rest; this is how Nadir started living as the son of soil. For trouser the army issued blanket was made good. Nadir hated the trouser but it was made compulsory to be worn by the officer in charge 'thus I would put

Subedar Nadir a veteran of 1948 War

it on for his pleasure and parade and would quickly change into shalwar'. During the operations the Indian Air Force attacked many times and it was in one such attack that two scouts embraced shahadat, ' we would rush towards the cover as soon as we heard the aircraft noise'.

Chitral-1948

Mehtar Muzafar ul Mlk died in 1948 and his son Saif ur Rehman who was already recognized as the Mehtar by the British India took over. It is worth noting that almost all the mehtars{Shuja ul Mulk, Amin ul Mulk} had the Afghan bloodline running from their mother side. Saif's mother herself was daughter of an Afghan refugee. Let it not be forgotten that all mehtars trace their lineage to Sangin Ali a Persian thus the Persians cultural influence is even today visible in the form of courtesies.

Chitral State was almost in chaos after the death of the Mehtar Muzafar, traditional plots to snatch the power were put into play. Mehtar Saif himself is partially responsible for the anarchy that ensued. In April 1949 the bulk of administration including the wazir-e-azam, commander in chief and chief secretary were all arrested and their properties confiscated. This was a real crisis for Pakistan as Afghanistan was always ready to exploit any weakness in the frontier states. By September 1949 the crisis were over with the intervention of political agent Malakand.

In **October 1949** crisis of another kind erupted when the marriage party from Dir which had come to take the bride; daughter of late Mehtar Nasir ul Mulk as per earlier agreement, were informed by the Mehtar Saif Ur Rahman that he himself has married the lady in June 1949 but did not disclose to the people; this set the stage for a civil war between the Dir and Chitral. Saif was later evacuated by air from Chitral and sent to Lahore to undergo training in civil services academy. It is worth mentioning that he had already received the military training from Pakistan Military Academy in 1948. The state was now put under a board of directors' with assistant political agent as head. Chitral at that time had governors in what are tehsils now. There was a governor at Drosh another at Mastuj, under the accession of 1947 the states retained their administrative set up as it was before accession. Saif later had a tragic end; he died in an air crash over Lowari in 1954 while he was coming to Chitral after years of wandering; all due to one fatal mistake of heart. He was succeeded by his four year old son Saif ul Mulk with Political Agent Malakand as the regent. This royalty was finally abolished by President Zulfiqar Bhutto in April 1972 when government abolished all such titles and privy purses. Just for historical fact the India had abolished all these princely states privileges under Indira Gandhi in 1971. Presently the Mehtar is a traditional title which people generally respect in the valley, Mehtar is presently living in the old fort on the banks of River Chitral.

Chitral Scouts -1956

In 1956 the Chitral State Scouts were reverted back to the identity of Chitral Scouts, for the reason that Pakistan became a republic from dominion status, all princely states were amalgamated into One Unit. Chitral Bodyguards also had their last breadth of short history; they were disbanded as the title of Mehtar itself was lost now. Chitral State Bodyguards and Chitral State Scouts were almost one identity, they had different financial regulations and bodyguards were exclusive affair of Mehtar. Amidst such chaotic environment the training and standards of the scouts were bound to drop. The scouts all belong to the same area and ethnicity; thus palace revolts had their impact on them. Gasht and long gasht were a valuable mean to keep the scouts busy. By 1956 they were organized into two wings known as **left and right wing**, the scouts were now a permanent military body. The normal gasht was of two days and long gasht spans over ten days. Each post was supposed to have eight gasht per month each ranging no less than 15 miles. Method of recruitment was simple, adjutant would inform the Wazir-e-Azam of the deficiency in manpower. Chief secretary was in fact known as Wazir e Azam , thus he would in turn inform or order the 'hakims' the village headmen , who would direct the desirous young boys for recruitment. The physically standards were; age no less than 17 years, height 5.6 inches, weight 120 pounds and chest expansion of 30-32 inches.. Recruits training duration extended to 36 weeks and conducted at Drosh. There was permanent corps drill staffs comprising of one jemadar adjutant, one corps havildar, three drill naiks and 5 drill lance naiks. For night training all ranks used to wear the boot made of skin{Kun}. At quarter guard the usual strength at day was two by six and at night two by twelve scouts. There were three quarter guards, one at Chitral , other at Drosh and one at Arandu. Drosh had an additional responsibility as well they daily had to provide a mail guard of two scouts for the protection of mail from Drosh – Mirkani.

For promotion a soldier had to be a first class shot. having a certificate of 3rd class education, should have passed the senior sepoy promotion examination and above all recommended by his wing commander, only then he was considered for promotion. Troops had a good ration , they were issued with meat thrice a week , they were entitled 11 chattaks of atta, half chattak of ghee and one chattak of daal daily. A special allowance known as khowar was admissible to the officers after they have passed the language, it valued at Rs 100 per month.

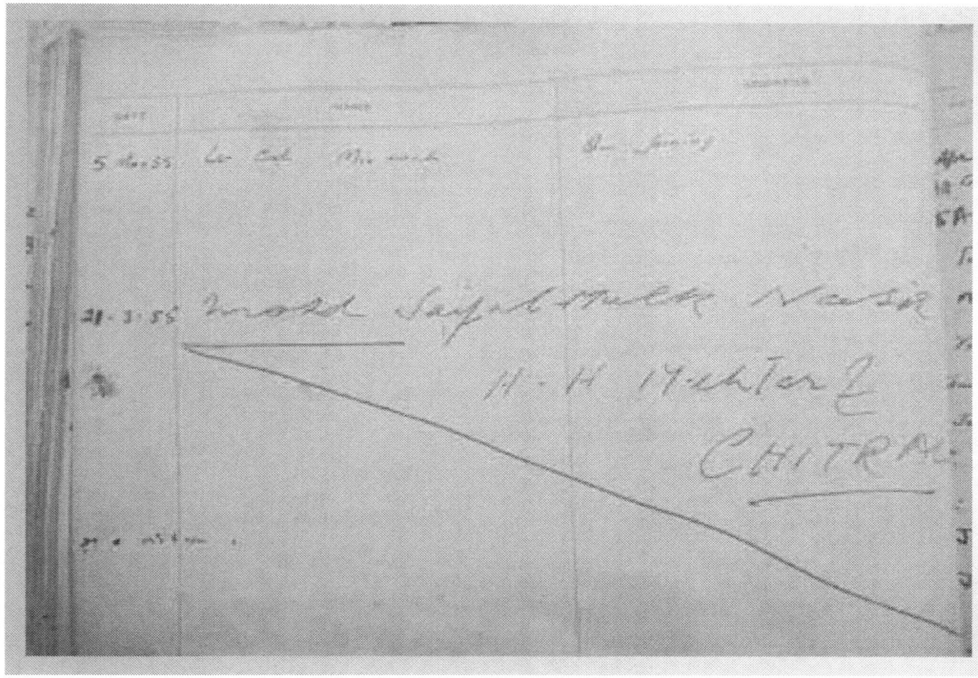

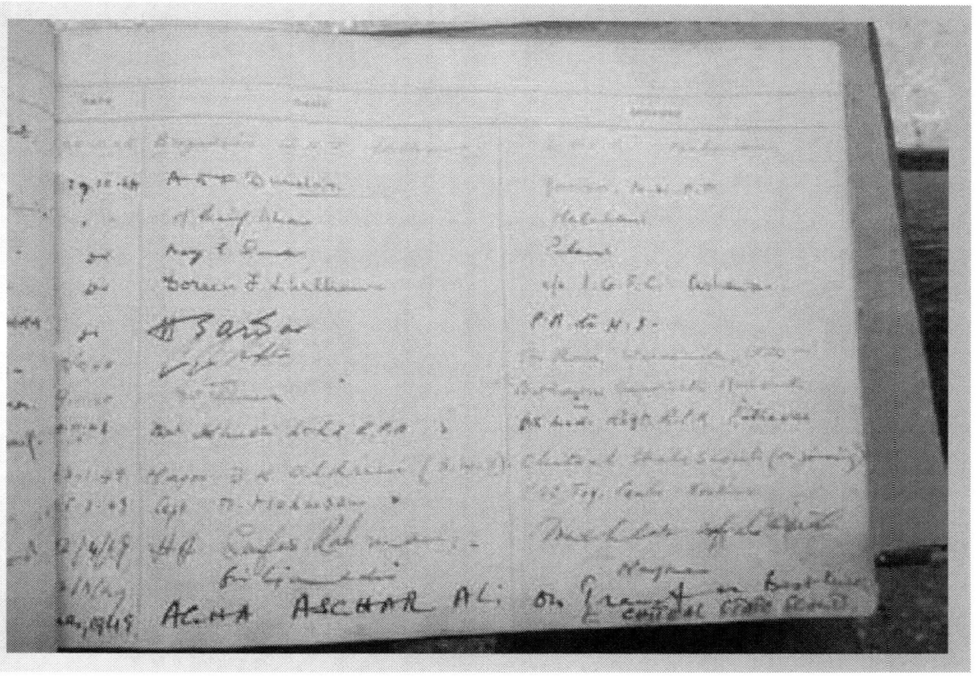

Chapter Four

Chitral Scouts 1965 - 1999

Chitral in sixties had a poor communication infra structure, the road between Dir and Chitral fit enough for three ton lorry was closed down from late November to mid June due to heavy snowfall on the Lowari Top. Only one ton lorry track was available from Chitral- Lowari. There was no road in the state, Mirkani-Dommel Nisar-Arandu, Chitral- Sanoghar, Chitral- Kiring, Chitral- Nagar and Gahiret-Barir were all jeep able tracks in fair weather only. The road Drosh- Madaglasht and Kiring – Garam Chashma were under construction. Chitral was connected with Peshawar by air, flights operating on fixed days. In January 1964 a new corps was raised at Skardu known as **Karakorum Scouts** with headquarters at Skardu, commanded by a major, it was raised to a lieutenant colonel's command on 20th September 1965[32]. At raising the half strength of the corps constituted from tribes of Baltistan and other half from Gilgit, Hunza, Nagar, Yasin, Punial and Astore; Chilasis and Gupis were recruited from September 1965 onwards.

Chitral Scouts were bit late in taking part in the 1965 War, the two companies of Chitral Scouts were ordered to move to the operational area in the second week of war and by the time these two companies reported at Rawalpindi on 23rd September 1965 the war was over. These two companies remained deployed at Rawalpindi where the threat of air borne troops from India was a distant reality. Subedar Suleman Khan was the commander of Chitral Scouts contingent, they remained at Rawalpindi till the last week of January 1966.

In the post 1965 War certain changes were introduced into the Chitral Scouts, army officers from services were also posted in like Captain Muhammad Yunis of Army Supply Corps who commanded a wing{No.2 Wing in 1966} Subedar Major Muhammad Zafar Ali Khan was granted honorary commission in the rank of lieutenant. Many of the old guards of 1948 Kashmir War bid farewell to arms in 1966 like Subedars Wazir Hayat, Kai Khesro, Aziz Muhammad and Zafar Ali Khan, Naib Subedars included Nazuk, Abdul Khaliq, Murad, Khurshid Alam, Mir Ahmad Khan, Muhammad Afzal, Sahib Nazar Baig, Abdullah Shah, Sahib Nadar, Rehmat Wali Shab and Jalal Sher; they all were part of Chitral State Scouts days. Many of the soldiers who were enrolled in 1948 now rose to the rank of subedar which included Subedar Hussain Ali Khan, Iltaf Ghafoor, Mir Bali Khan, Zain ul Abidin, Fida Muhammad Khan and Muhammad Zahir Khan. Sports were played with festive and competitive spirit; it was on intercompany basis. In 1966 Intercompany football was won by A Company of No.1 Wing, Basketball by Headquarters Company and shooting by D Company of No.2 Wing , most prestigious the Khud Race was won by C Company of No.2 Wing.

Northern Scouts- Frontier Corps Northern Areas.

In 1949 the Gilgit Scouts after the liberation of Gilgit were reverted back to their normal duty of internal peace keeping under the political agent and a new force known as Northern Scouts were raised on 1st November 1949 for the defence of cease fire line{CFL}. A deputy inspector general appointment was also created who was responsible for the northern areas. In January 1966 Colonel Haq Nawaz assumed the command of

32 *Frontier Corps an Introduction*, p-191.

FCNA vice Colonel Pir Abdullah Shah who retired. There was one administrative officer {Major Liaqat Ali Khan and Major Azam in 1965-66} and one officer {Captain Muhammad Arshad in 1966} who acted as the officer in charge of despatches at Chaklala. FCNA comprised of Northern Scouts and Karakorum Scouts. In 1967 the population of Gilgit Agency including Hunza & Nagar was 1,67,245.[33]

Chitral Scouts Officers with Commandant Lt Col Zia ul Haq (Sitting 3rd from right) 1963

In 1973 , FCNA was made into Force Command Northern Areas a nomenclature that still exist today. The scouts were upgraded into Northern Light Infantry battalions. Despite all the passage of years the basic administrative infrastructure introduced by the Frontier Corps in abinitio days is still in motion. From 1992 -2008 , Chitral Scouts served under FCNA in one capacity or the other .

1971 War.

The situation in the former East Pakistan was getting deteriorated every day and new raisings were ordered by the government. In November 1971 the Commandant Lieutenant Colonel Afzal in a darbar updated the scouts with the situation and asked for volunteers , four scouts namely Havildar Sarfraz Shah alias Khamshay of Laspur, Naik Nasir Shah Bang{Yarkhun}, Naik Qadeer Khan Raman{Laspur} and Naik Khush Ahmed Khan {Mulkhow} stood up and volunteered themselves[34]. Commandant in recognition of their patriotism promoted all these scouts to next rank before their move to the headquarters Frontier Corps. All these four sons of Chitral Scouts were later made prisoner of war after the fall of Dacca on 16th December 1971. They were repatriated under the Delhi agreement {Simla} of 18th August 1973 but were able to join back with their families in December 1974. Among these Naib Subedar Sarfraz Shah Khamshay was able to air his interview through All India Radio in which he highlighted that they are being treated well and according to the Geneva Convention' I am fine and life inside the prison is also fine'. In the end he remarked in Khowar language '***Gooshkoo Moraien Zannan Soureein***' which is a Chitral proverb meaning 'everything said and heard is false'.

33 *Frontier Corps an Introduction*, p-137.
34 Naik Khusro, *Chitral Scouts*, p-55

Eid Function - 1971

Pakistan Army plan in Kashmir during 1971 War necessitated reduction of troops in Lipa Valley with an intention of launching attack from Poonch Sector. More than 140 scouts from Chitral Scouts took part in the 1971 War on Western front at Kargil Sector. They remained at Kargil for over two years, another batch of 40 scouts performed duty at Rawalpindi garrison.

Chitral Scouts 1973 - 1992, Two Forgettable Days

Chitral Scouts in 1973 and in 1992 had two bad forgettable days. In 1973 while at Drosh, with Lieutenant Colonel Afzal in chair { he wrote the very first post 1947 booklet on Chitral}. Frontier Corps was in an expansion phase with Brigadier Naser Ullah Babar in command. It was decided in principle to post the Chitral Scouts platoons to other corps also and similarly to induct other qaum in the Chitral Scouts. Chitral Scouts resented it and took up a defiant stance led by Subedar Major. Finally despite commandant's orders they left Drosh and fortified themselves at Chitral. Brigadier Babar flew in a helicopter but Chitral Scouts rebelled segment watered the polo ground there by making landing rather difficult for Brigadier Babar. In the end a compromise was reached with Brigadier Babar, Mehtar and subedar major as signatories, under which the scouts agreed to serve outside the Chitral in the form of a complete wing rather than in platoons. Later one wing of Chitral Scouts remained deployed at North Waziristan Agency for over a year.

Other incident in 1992 was a result of misunderstanding, the Chitral Scouts contingent had gone to Nowshehra , the senior subedar in the evening parade state did not show the missing vehicles which he had sent to the Mardan on private job. In the end the contingent was recalled and over 64 scouts were charge sheeted and put on trial. Almost all were given punishment by the Major General Ghazi Ud Din Rana, sepoys were given sentences ranging from seven days to twenty eight days in cell, rest all were demoted with senior most being given imprisonment in civil jail. Reason for such mass scale punishment was in the fact that all scouts, being from one qaum took the side of subedar less four who came out with truth. In 1996 Major General Naseer Ullah Babar {retired} now the interior minister took a lenient stance on mercy appeals of convicts and restored their pensions.

Prime Minister Bhutto at Chitral 1973

Working Party Chitral Scouts 1970

Drosh Fort 1972

Governor Arbab Sikandar 3rd from Left

Gasht in Kafir Valley 1972

Governor Ghawas at Drosh

Chitral Scounts Band

Chitral Mosque

Drosh Mosque

Grave of Lt Col Nayer

Commandant House

President General Zia ul Haq with
Lieutenant Colonel Murad Nayyar behind
him, reviewing the Chitral Scouts Guard of
Honour at Shandur 1984

Now these are rare scenes and unprecedented in nature. It only highlights how important is the subedar major and how vital it is for the officer to keep an eye on the pulse of the things, both events reminds of Gilgit Scouts of 1947, almost a replica. That is another dimension of these acts; they have occurred only in non Pashtu high altitude mountainous region, which were both former princely states, they have only one qaum. However the Chitral Scouts & Gilgit Scouts mutinies have always been non violent.

Lieutenant Colonel Murad Khan Nayer.

History of Chitral Scouts and Chitral itself will be incomplete without mentioning the name of Lieutenant Colonel Murad Khan Nayer. He remained commandant of Chitral Scouts for twelve long years from 1977-1989; he was the uncrowned Mehtar of Chitral. His legacy includes two fantastic, magnificent mosques, one at Drosh and more spectacular at Chitral. He inaugurated Shandur Polo tournament to present glamour {he himself was not a Polo player}. Murad was an old hand of Frontier Corps, he had served in Kurram Militia in 1967, as a staff officer in Frontier Corps at Balahisar before assuming the command of Chitral Scouts. He made Chitral Scouts a champion corps winning championship banner thrice and always a medal in annual sports competitions. He harnessed the natives in the Chitral during Russian occupation of Afghanistan; it is pure coincidence that both ended simultaneously. A poet of enjoyable words, patron of arts, music & craft in Chitral. He was among the closest friend of President General Zia Ul Haq; although Murad was much junior to him in service. It was in 1967 when both met for first time, Zia then a lieutenant colonel was moving with his family to Kurram for a brief stay and it was here that Murad became his host. Zia and his family was much impressed with Murad's hospitality and sincerity. Legend goes that Murad prophesized to Zia that he had seen in his dream of him becoming the army chief ; Zia just laughed it out but then agreed to accepting whatever Murad wishes in case if he becomes an army chief. General Zia not only became army chief but also the President of Pakistan; he stood by his promise and kept on giving unprecedented extensions in service to Lieutenant Colonel Murad.

General Zia had his last breath in an air crash in August 1988 and next year Lieutenant Colonel Murad committed suicide while still being the commandant of Chitral Scouts. He took his life in the mid night hours in his house, Murad was a bachelor; he left a note also highlighting that he is taking his life with his own will and none from Chitral Scouts should be held responsible for this act. As a last will he requested to be buried in Drosh. Presently he has attained a status of a saint among the natives who remembers his kindness and generosity towards the natives especially poor. His marble grave is in one corner of the Drosh mosque which he constructed. The house he occupied for twelve years at Drosh has been demolished to make room for the Chitral Scouts School. People still believes that his ghost is living inside Drosh Killa and on full moon takes a round of killa.

Kargil 1999

Chitral Scouts were once again in Kashmir in 1999 during Kargil War which itself was a continuation of the 1947 Kashmir War. The natives had risen against the Indian occupied troops and once again like 1947 they were helped by the tribes and local population. Mujahedeen soon captured the important dominating heights overlooking the sole highway linking Kargil with Leh. In the summer of 1999 the war clouds loom high and dark over sub continent with both countries carrying out nuclear explosion as well.

SPECIAL PART-I ORDER
BY
PA-16173 COLONEL UMAR HAYAT KHAN COMDT CHITRAL SCOUTS
CONDOLENCE
WEDNESDAY 16 JUNE 1999 CHITRAL
Commandant and all ranks of Chitral Scouts are deeply shocked and grieved on the demise of following officers/OR of this unit who embraced Shahadat-e-Uzma on 16 June 1999 at about 1530 hrs due to Enemy bombing while defending mother land at Rader Post in Hamzighund sector under FCNA :-
a PA-26773 Major Muhammad Ali Hyderi, 30 Baloch
b PA-32762 Capt Zulfiqar Ali, 110 Fd Regt Arty
c 402838 Sep Abdul Wahad, Chitral Scouts
The fol received serious injuries :-
a 401415 Lnk Shafiur Rehman, Chitral Scouts
b 403727 Sep Zaffar Ali Khan, Chitral Scouts
"TO ALLAH WE BELONG AND TO HIM IS OUR RETURN"
May ALMIGHTY ALLAH rest the departed souls in peace and grant strength and courage to the bereaved families to bear irreparable loss with patience and fortitude. "AAMEEN". May Allah Almighty Allah also recover the injured soldier so that they could join the combatants in Arm the fight against the Enemy.
Capt
OC Adm Int
(Muhammad Nasrullah Khan Janjua)
Address for Condolence
1 Asma Ali w/o Major Muhammad Ali Hyderi Shaheed 74 Arif Jan Road Lahore Cantt.
2. Parveen Zulfiqar w/o Capt Zulfiqar Ali Shaheed H No 7-22/91 Qadam
Capt Zulfiqar
Maj Muhammad Ali
SHUHADA-E-KARGIL
CHITRAL SCOUTS 1999
1. NK Munir Ahmad 2 Aug 99
12. L/NK Sikandar Khan
13. Sep M. Jalal-Ud-din
4. Sep Abdur Rehman 26 Jul 99
YAADGAR-E-SHUHA
THIS MONOMENT HAS BEEN
IN THE MEMORY OF IMMORTAL SAC
OF GALLANT SOLDIERS OF CHIT
SCOUTS WHO MOVED FROM MIRKHANI
KARGIL SECTOR ON SHORT NOTICE UN
THE ABLE COMMAND OF MAJOR ALI HY
SHAHEED THE TROOPS PUT THE EN
TO ROUTE BY A MORTIFYING DEFEAT TH
SUPREME SACRIFICES SHALL BE REMEMB
WIHT VENERATE IN DAYS TO COME MIR K
FORT WAS RENAMED ALI SHAHEED FORT
RECOGNITION OF HIS SUPREME SACRIFICE
INAUGURATED BY
MRS MAJ MUHAMMAD ALI (SHAHEE
YAADGAR - E - SHUHADA KARGIL
WE DESIRE DEATH
MORE ARDENTLY THAN
YOU DESIRE LIFE
YAADGAR - E - SHUHADA KARGIL
MAJOR MUHAMMAD ALI HYDERI SEP SHAHBAZ KHAN SEP ASHRAF NAWAZ
CAPT ZULFIQAR ALI SEP ABDUL WAHID SEP MEHBOOB KHAN
NK ASHRAF KHAN SEP ALI HAZRAT SEP FAZAL KARIM SHAH
NK MUNIR AHMAD SEP NOOR KHAN SEP ABADUR REHMAN
SIKANDAR KHAN SEP MUHAMMAD JALAL UDDIN
SHOHADA RASHID SECTOR SKARDU
L/NK. Shakir Ud din 5 Jan 2004
Sep Nasir Ullah 5 Jan 2004
L/NK Muhammad Shafi 27 May 2005
Sep Samiud din 27 May 2005
Sep Shujaul Islam 27 May 2005

First one to enter into the arena was **No.2 Wing** which reached Skardu after four days of travelling on 28th May 1999, after a day's rest it moved to Shaqma and Hamzigund. Wing headquarter was placed under command Northern Light Infantry and a company was placed in Idress Sector. Wing moved further ahead to the Ganokh and then to Imtiaz Position. Wing's major task was portering of defensive stores including ration to forward posts, wing also had to construct defensive work for own protection; this continued till 1st June 1999. From this date onward Indians started extensive artillery shelling and air strikes were also called by them. On 16th June 1999 during one such air strike Major Muhammad Ali Hyderi and Captain Zulfiqar Ali embraced shahadat. Major Waleem Wali was rushed from Gilgit to take over the command of the wing. On 13th July 1999 cease fire was agreed upon by both countries. However that was not the end of hostilities, portering was carried out by the wing, it had two companies at Imtiaz, one company at Ganokh and one company at Hamzigund. On 7th September 1999 Lieutenant Colonel Abdul Qadoos took the command of the wing but he was relieved shortly by Lieutenant Colonel Abdul Ghafoor in December 1999. In June 2000, No.2 Wing was relieved by **No.3 Wing** of Chitral Scouts. On 19th June 1999, **No.1 Wing** of Chitral Scouts under the command of Lieutenant Colonel Kaleem Ud Din reported at Rawalpindi , they were ordered to camp at Partuk near Piun which involves almost 700 kilometres of journey. They arrived at Partuk on 24th June and received operational orders from the brigade 'Detach one company each with Azad Kashmir and Northern Light Infantry, One Platoon for the protection of Surmo Bridge and remaining wing to form part of brigade reserves'. Fluid situation resulted in fresh orders under which Wing established no less than eleven posts in Kusting Nullah sector, Captains Zulfiqar, Aijaz Hussain, Ali Mehar, Naib Subedar Sawada Jan and Subedar Sardar Wali played key role in the establishment of posts, major bases, mortar position, tactical headquarters at heights which vary from 13,600 -17900 feet.

Kusting Nullah a glaciated open area of four kilometres length which links own two brigades. It was manned by a platoon of Baluch regiment, now the Chitral Scouts were responsible for the sector. Chitralis soon developed it into a defensive location. In the process the first contact with Indians was made on 14th July. They immediately started shelling on Chitarlis. There was lack of defensive stores yet no serious damage was caused by erratic Indian artillery. All in all 4971 rounds were fired by the Indians which resulted in the injury of two scouts. Anab-1 received 443 rounds, Akbar OP received 656 rounds, Sardar Mortar got 590 rounds. In retaliation own artillery replied befittingly, artillery heavily depends upon correct observation and correction and for this observer has to be not only good but bold too. Enemy positions at Twin Peak, Flat Ground and Knoll Observation Post were hit hard; in fact Knoll OP was destroyed twice by Captain Aijaz's observation. Forcing enemy to take up defense on reverse slope. Subedar Sardar Wali was commanding the Mortars and he was quite effective in it, 520 rounds of 82mm Mortars were fired along with 52 rounds of 60mm Mortars. Sector then became quiet after 20th August 1999. Between, 1-10 September Indians again tried to infiltrate the sector but failed. The three detached companies of the wing also lived up to the reputation; they established three new posts known as, S-1, S-2 & S-3 at 17600 feet by Naib Subedar Abdul Aziz, {S-3 at 17900 feet by Naib Subedar Muhammad Sharif.}. Portering is the key to success in glaciated warfare and it requires stamina and body muscles which are tuned to mountain climb. Chitral Scouts dumped 41,600 kilograms of dry ration, 46100litres of K-2 oil, 80,340 rounds of G-3, 248484 rounds of light machine gun, 700 mines, 1094 grenades and above all 434 rounds of artillery in Kusting Nullah sector alone. Detached companies also matched the efforts. No.1 Wing Chitral Scouts also constructed 15 kilometres long mule track linking Akmal gun position to mid point.

No.1 Wing lost four scouts who embraced shahadat they were Naik Munir, Lance Naiks Sikander & Ali Murad and Sepoy Jalal. Mention in despatches. Major Saqib Zarin, Captains, Najam Ul Zaman, Mansoor Zeeshan, Aijaz, Zulfiqar, Mehar. Subedar Major Ghulam Rasool, Subedars included Mir Nawab, Ali Zafar, Ibrahim Wali, Mir Habib, Haji Murad, Shahid Zarin & Sher Rehman. Naib Subedars Muhammad Nasir, Sultan Ghazi, Sawada Jan, Qurban Ali, Dinyar Khan and Abdul Aziz. Havildar Iqbal Shah,. Naiks Munir Ahmed, Noor Ahmed, Rehmat Khan. Lance Naiks, Sikander Khan, Ali Murad. Sepoys , Sher Ghulab, Feroz Kahn and Rehmat Ali Shah.

Reminiscences of Kargil

Sepoy Rehmat Khan recalls ' *On 18th May 1999 , No.2 Wing was ordered to move to Kargil, we were given tremendous send off by the town of Drosh, almost the complete city was on streets cheering us, we were garlanded, after three days of journey we reached Skardu. Our arrival and arrival of six dead bodies of NLI coincided. Next day we again travelled and reached our destination, next day sector commander had a Durbar and he briefed us on the current situation, he also inquired that if anybody has any family problem he can highlight it now , but all scouts were volunteered for the action. Wing was being commanded by Major Muhammad Ali and Captain Zulfqar was second in command, subedar major was Jamil Ahmed. On 22nd May 1999 we move from Skardu and by 2200 hours were at Hamzigund. Next day on 23rd May we were ordered to move forward but on our way enemy artillery was intense and we were ordered to revert to our old position. On 24th May at midnight we moved forward again and reached Ganokh; half an hour after our arrival we were shelled by Indian artillery, it killed one civilian woman and a child and one was injured. On 25th May we were issued with snow line equipment and uniform, we did not have our breakfast and moved ahead as we had to relieve one company. Our one company under Subedar Rehmat Nazir and Subedar Ghulab took us to Imtiaz Post where we established our company headquarters, we had nothing to eat. On this night we received 102 artillery shells from enemy without causing any damage. After two days we were ordered to move again and we trekked for five hours before we reached our destination, it was a walk in deep snow. Soon we were ordered to move further ahead and through a narrow gorge using ropes we climbed up and then descended down, it was a plain piece of ground , snow tents pitched with a NLI unit manning it. Night was very cold. Next day we were split into five men section and attached with NLI troops. We were also given the task of portering the ration and ammunition, it was very tough . We used to do it at night. Our post was called as Badr Post, there were six or seven more posts ahead of it also, we had the responsibility of taking up defense of these alongwith NLI troops. We were under continuous enemy fire attack.*

On 15th June, our Wing Commander Major Muhammad Ali visited us at evening and spent the night with us. Next morning at 0840 hours two enemy jets pounded our position but none of the bomb hit the base camp. At 0923 hours again enemy jets attacked our camp, it was intense, the sound, shrilling noise, explosions, we all took cover behind the boulders. It was during this attack that we lost our two officers and three officers of NLI along with six soldiers {Sepoy Abdul Wadood martyred and Lance Naik Shafi was wounded}. Enemy remained in air for over two hours and after the sky was clear , we started searching for our comrades dead bodies. It was very tough time because our wing commander and the officer had embraced shahadat in front of us. Instead of being demoralized it further raised our morale, we all shouted Allah o Akbar , it was so full of energy that it echoed through the mountains. Bombing had turned the whole area into slush. Soon own helicopters came and evacuated the wounded first and later they helilifted the dead bodies also which were brought into rear by foot. On 26th June 1999 we were again hit by the enemy air and Naik Ashraf Khan, Lance Naik Ashraf Nawaz, Sepoy Fazl Karim, Sepoy Noor Khan, Hazrat Ali and Shahbaz embraced shahadat.

We were receiving letters from our home which was a great morale booster; the people of Northern Areas show their love by sending us gifts and edibles. We left Kargil in August 1999 for journey back home, we were again given a tumultuous reception by the town of Drosh and where ever we went '.

Chitral Scouts Quarterguard at Chitral Fort with captured Afghan Gun of 1919 War. Fort itself was constructed in 1895.

Chapter Five

Chitral Scouts Forts & Messes - 1895 – 2014

Chitral Fort is the oldest among all forts of Chitral District. It was constructed in 1895 in the aftermath of Chitral siege. This fort was constructed two miles down stream & south east of the Mehtar's fort and old residency thus it has more commanding position. It is the only fort which British constructed on the right bank of River Chitral. Construction is simple, citadel in pattern, four corners with piquets, solid stone bricks, a twenty feet high wall all around encompassing almost half a kilometers of length. It is single storied and relatively flat in nature. There were two cordons of construction, the outer being used for living and inner for magazines. One main gate facing the River Chitral which soon after entrance takes right angles turn to break down momentum. The other exit was on eastern end, opening into the fields stretching close to the bank of river. It is presently used as the sports, heliport and polo ground. The Chitral Scouts Public School is also located at the far eastern edge of the fort's ground. On the southern and south-western side the area is open, broken, populated and vegetated, in the past it was rather isolated and desolated. A hanging bridge also connected the fort across the gol to the Mehtar's fort and bazaar area, the bridge is still intact and new road is under construction to link the Fort with bazaar. In September 2013 the bazaar was demolished to expand the road thus a century old tradition came to an end.

In 1899 an independent suspension bridge over River Chitral was constructed exclusively for the fort[35]. In 1985 a beautiful mosque was constructed adjacent to the green bank of the river by Lieutenant Colonel Murad. The bridge remained in use for over a century and only recently has been closed down. Chitral Fort over period of time has become the focal point of all events and visits in Chitral. With the shifting of landing ground from Drosh to Chitral in 1960 the town started emerging as the capital of the state. In 1992 the Chitral Scouts headquarters were also shifted back from Drosh after almost seventy years and they now occupied the present day fort, later in 1999 the hospital also moved here. With these expansions the original walls of fort have undergone modification; commandant's house is also adjacent to the fort so is the medical staff accommodation. Mounted infantry platoon, signal platoon, military transport platoon all resides here, fort is much smaller than the Drosh Fort. At Chitral initially the officers mess was inside the fort but in 1934 the Mehtar offered his summer palace grounds to the Chitral Scouts Officers where present day mess is now situated. It takes almost ten minutes of drive to reach fort from mess and in winter when the road is covered with feet of

A view 1971

35 Captain James of Royal Engineers Report 1899, Chitral Scouts Archives, Chitral

Chitral Fort

Chitral Fort during Winter

Chitral Fort 1895

Chitral Scouts Mosque at Chitral, built in 1989

Promotion is an occasion which is celebrated with much fanfare and add colour to the Fort Life

snow the driving becomes a test of skills. All troops livings are inside the fort where new barracks have been constructed, junior commissioned officers mess is also located inside the fort. Canteen and other services like tailor and cobblers are also working inside the fort. Present day quarterguard and offices area were constructed to accommodate the shifting of headquarters from Drosh and as such are the first expansion of the old fort. Fort environs have graceful, fantastic, mature and grand trees; it is the cultural hub of Chitral. Daily over one hundred civilians visit Chitral Scouts hospital and over five hundred pupils come daily for studies; both institutions are the best in Chitral, not to mention the cafeteria and bakery of the parks. Life in the fort is regular and disciplined, wake up at Fajr time, breakfast and then routine work. Nature of work is fluid especially in summer and in present situation. Guards and other sentry posts are alert and manned fully and with this protection the other routine works carried out as usual. Officers arrive from the mess by vehicle and they seldom leave by afternoon . Evenings are special in the fort as sports are played with zeal and enthusiasm, more to relax mentally. Cricket matches are regular feature with headquarters and medical staff being two rivals, the limited over matches lasting well into darkness every evening. On 3rd September 1999 Chitral Scouts Hospital was established at Drosh and on 17 December 1999 this hospital was shifted to Chitral (on its present location) General Musharraf Chief of Army Staff inaugurated it.

Petako Gaz-1934 - **Chitral Scouts Officers Mess.**

The present day Chitral Scouts Officers Mess is known as Petako Gaz in Chitrali language , it has a history, many many years ago there was a Khonza {the royal princess} who came here and forgot her dupaatta here and when she went back to the palace she remembered her cloak and mentioned it to her attendants and gave the direction to the place as well, in Chitrali language Petako Gaz means a place where princess forgot her veil. The present location was at one time part of Mehtar's property and even now the Chitral National Gol Park almost starts from where the mess finishes.

Petako Gaz is in almost seven layers or tiers of ground which have been levelled enough to construct the building. The original building of the mess was constructed in 1934 and from then till 1990 it remained in its original shape for the reason that Chitral Scouts itself moved out from here and established headquarters at Drosh and it was only in 1995 that they came back to this Mess again. Many additions have been carried out but the original design has been left intact, credit goes to the commandants for maintaining the original layout which is more or less like Swiss dacha.

The most notorious stone on frontier is here the 'Laying stone of Captain Boono'. Mess presently have apart from two original guestrooms known as Birir and Rumbor have almost a dozen other guest rooms and living quarters for the adjutant and one wing commander, accommodation for mess staff, signal detachment and security guards. There is a diesel generator, one hydel power which was commissioned in August 2013, gymnasium, library and billiard room. One clay tennis court at tier two and one squash court at tier three. There is one mosque as well.

Mess have two main halls which are rectangular in design , one is used as dining room and other as ante room having television facility; in old days there used to be a piano and radio followed by radiogram.

In 2002 the guest room charges were rupees 200 per night for lieutenant colonels and above and rupees 150/ night for all others while on leave at Chitral Mess. Mess in order to maintain its decorum usually restricts the entrance of children under twelve on official functions. Chitral Scouts have one peculiar issue almost all their official inspection and tour takes place in summer time mainly due to blocking of road movement in winter thus they have heavy load of guests in summer therefore it is always advisable to have an advance booking here.

Painting of the Chitral Scouts Officers Mess, 1935

YOU ARE AT
E = 735448. 26
N = 13088. 25
HT = 5247. 12
VY BY MAJ BABAR (LOCATER)
1982

Chitral Scouts Officers Mess 'Petako Gaz'

Library
Mess Staff

COAS DINNER IN CHITRAL SCOUTS OFFRS MESS
3 SEP 1999

General Zia-ul-Haq Fourth from Left at Petakogaz 1982

Chitral Scouts Officers Mess Silver

SCOUTS
BY
KHAN

Drosh

CHITRAL
GAHIRET
BUMBURET
TEH DROSH
DROSH
URTSUN
MIRKHANI
D. NISAR
ZIARAT
LBT
ARANDU
LAWARI PASS
GAURDESH

Drosh Fort; 1899-2014

There are two forts in Drosh one is known as Chitarli Fort and other as Drosh Killa, the former has been turned into a school as back as 1937 while other remained as headquarters of Chitral Scouts for almost fifty years till 1992. There are two piquets on high ground behind the fort, one is on the north and other is east of fort. The fort is large, big, magnanimous and simple in construction, it is in tiers and have solid walls which were broken are replaced with barbed wires. Originally there was only one entrance into the fort on the western side having a piquet it is disused now; a pity. There are two rings of fort , one which is uncovered by wall and other the proper fort where the gate leads you to the quarter guard and then into the fort. Presently there is another path which is for motors which can lead up to mess but one still has to climb stairs to reach to the mess. From quarter guard another inclined flight of stairs leads to the commandant's office and from there to the mess. It is at quarter guard that an original wall painting has been preserved although refreshed every now and then which is a mirror of the past. It shows that almost everything was on the northern bank of river and it belongs to Chitral Scouts and the assistant political agent used to sit here instead of Chitral.

Drosh Fort's construction started in May 1899 and was completed in October 1900. The fort is on the northern bank of River Chitral, on a higher ridge overlooking the complete valley and route of River Chitral and River Madaglasht. Thus from its siting it is obvious that main threat was from the Afghans and the southern circle was the most volatile. In the ancient times the track from Bashgal in Afghanistan entered into Chitral through Arandu and then following the hill track it passes through Dommel Nisar-Mirkhani-Drosh-Ayun-Chitral-Garm Chasma and then exiting through Shah Salim Pass. Thus two separate entities one on the southern bank and other on northern bank were flourishing. There were very few crossing points on the river and even then the hanging bridges were unable to take the full load of laden camel, horse or mule caravan, only in military expedition a force was able to cross it. In summer the river was in full fury as now and in winter the crossing was possible.

Drosh Fort is built on the ridge in a multi tier design, which is inclined, and hardly any level space is available and even less was in the past. Fort is rectangular in design having mud barracks all along the four walls; these barracks are the distinct hallmark of the fort. The wooden pier style corridor along walls are narrow yet wide enough to allow two men to cross each other. Firing points are available after every seven feet; the best part of the fort was in its rapid concentration of fire and soldiers at all time mainly due to the design. Four piquet at four corners along with equal number of small entrances built of iron doors and one main entrance, which in the past had a gigantic iron and wood door. Quarter guard is on the right side of main gate inside the fort.

First major expansion of the fort took place in 1921 after the Third Afghan War when new barracks were constructed and headquarters of Chitral Scouts was shifted from Chitral to Drosh. Second expansion took place in 1939 and then in mid eighties it was almost complete in every sense when the headquarters were moved back to Chitral. Since then the fort is living in past nostalgia as lack of manpower is barely enough to keep it in working state, in 2011, Lt Col Salman carried out the maintenance and renovation.

In the original pattern the complete northern bank of River Chitral was only with the military and political administration. The road leading from fort to the bridge on the River Chitral was and still is the property of Chitral Scouts and that include the complete bazaar with over two hundred

DROSH
FORT
BEGUN
MAY 1899
COMPLETED
OCT^{R} 1900

DROSH CEMETERY
FOURTEEN PERSONS EX BRITISH ARMY HAVE BEEN BURIED IN THIS CEMETERY FROM 1896 TO 1935 DETAILS ARE AS UNDER
RANK NAME REGT DATE OF DEATH REMARKS
CAPT JAMES EADS PRINCE ALBRT 13-10-1896 DIED WHILST EMP AS TO AT DROSH
COL CLAUDE MELVILLE GORKHA RIFLE 26-6-1912
LIEUT ERNEST ALBERT ROYAL CARNIVAL
LIEUT DENNIS NEVILLE ROYAL ENGR 15-10-1935 DROWNED IN CHITRAL RIVER
FRANK SPENCER R-A-F 15-9-1930 KILLED IN FLYING ACROSS
SERGEANT OWEN CAMBRIDGE R-A-F 15-9-1930
NO 4202 SAPPER CRUZE 2ND Q-V-O-S&M 27-9-1915
HDP 4778 Q-V-O-S&M 25-11-1918
DETAILS OF MUSLIMS BURIED IN THE GRAVEYARD ARE AS UNDER
JAMADAR FARHAN ALI 4/6 RR 1928
L/NK SAKHI MUHAMMAD RR 15-2-1930
SEP QURBAN ALI PR
BOSTAN RR 17-2-1930
ZARDAT KHAN RR
GAZMIR HAYAT 15-11-1935
AKRAM HAYAT SOO DROSH 3-2-1947
DONE BY 42 AK
(HAT BN)
TANK KILLERS
OCTOBER 1905
DROWNED IN CHITRAL RIVER
TRYING TO SAVE ANOTHER'S LIFE.

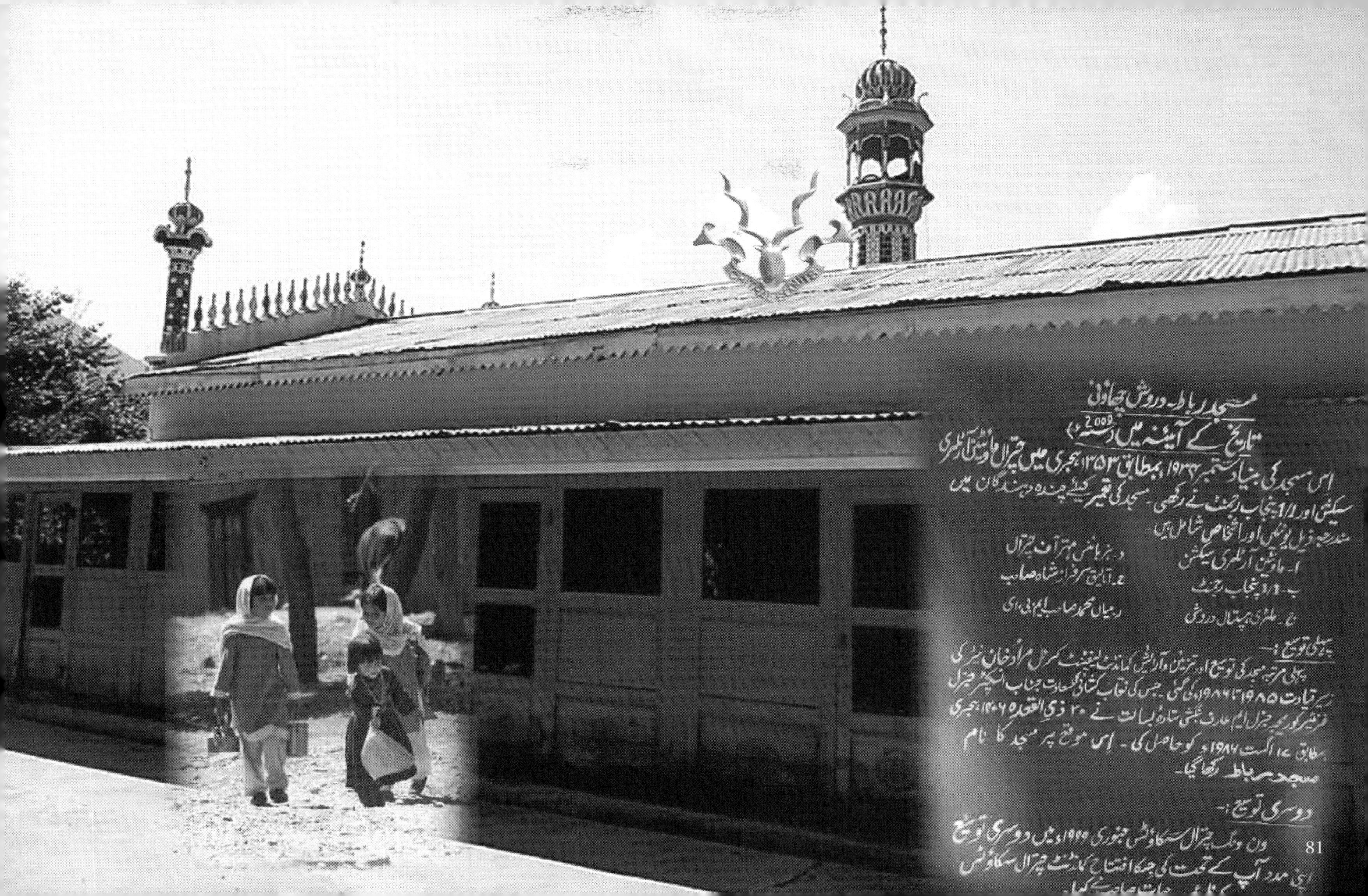
مسجد رباط۔ دروش چھاؤنی
تاریخ کے آئینہ میں (دسمبر 2009ء)
اس مسجد کی بنیاد ستمبر ۱۹۳۴ء بمطابق ۱۳۵۳ ہجری میں چترال ماؤنٹین آرٹلری
سیکشن اور 1/1 پنجاب رجمنٹ نے رکھی۔ مسجد کی تعمیر کیلئے چندہ دہندگان میں
مندرجہ ذیل یونٹس اور اشخاص شامل ہیں۔
ا۔ ماؤنٹین آرٹلری سیکشن
ب۔ 1/1 پنجاب رجمنٹ
ج۔ ملٹری ہسپتال دروش
د۔ ہز ہائینس مہتر آف چترال
ہ۔ اتالیق سرفراز شاہ صاحب
ر۔ میاں محمد صاحب ایم بی ای
پہلی توسیع :۔
پہلی مرتبہ مسجد کی توسیع اور تزئین و آرائش کمانڈنٹ لیفٹیننٹ کرنل مراد خان نیئر کی
زیر قیادت ۱۹۸۵ تا ۱۹۸۶ء کی گئی۔ جس کی نقاب کشائی بسعادت جناب انسپکٹر جنرل
فرنٹیئر کور میجر جنرل ایم عارف بنگش ستارۂ بسالت نے ۲۰ ذی القعدہ ۱۴۰۶ ہجری
بمطابق ۱۷ اگست ۱۹۸۶ء کو حاصل کی۔ اس موقع پر مسجد کا نام
مسجد رباط رکھا گیا۔
دوسری توسیع :۔
ون ونگ چترال سکاؤٹس جنوری ۱۹۹۹ء میں دوسری توسیع
اپنی مدد آپ کے تحت کی جسکا افتتاح کمانڈنٹ چترال سکاؤٹس

Drosh Officers Mess

Drosh Mess Staff

Drosh Officers Mess Silver

shops. Now the bazaar which has sprung up along the main road is known as new bazaar and Scouts Bazaar is called the old bazaar.

Initially the commandant house was inside the fort but in 1920 it was constructed outside the fort and then finally demolished in 2010 to make room for the expansion of Scouts School. Chitral Scouts garden is located across the river, this is one of the largest garden among all corps of Frontier Corps, the old landing ground which was built in 1921, last had flight in 1960 and since then is abandoned and turned into a cricket field with proper brown pitch, it is a fantastic field.

Fort had three grave yards, one for each religion, one known as cemetery have both the Muslim and Christian military personnel's and is located at the south eastern end, the Hindu Shamshan Gath was situated a mile further east but now a days there are no remnants of that graveyard, however the other cemetery is maintained by the scouts and the army unit stationed in Drosh {they are here since 2009}.

Presently the contractors bring fresh ration, fodder for animals, petrol and other commodities as they have been doing it for last hundred years. The Chitral Mountain Artillery is also stationed here, it has been its home base for a century and despite all other changes this has remained in vogue. Mounted infantry also originally had their birth here, the stables are still functional and the horses, mules, donkeys all live together, {mounted infantry only looks after the horses and animal transport is responsible for donkeys}.

Presently the lush green hockey field serves as football ground with daily matches being played between the asr-maghrib prayers, it is a treat to the eyes to see such colourful attires running around ball. In the parade ground, which is now known as Murad Ground the civilians are allowed to play the games in the evening, as there are no other sports facilities available to them, similar is the attitude towards the use of cricket field.

Mountain Hospital was also at Drosh before it was also shifted to the Chitral. The present half mounting barracks were once the hospital wards, there are two such barracks, the bigger one was general ward and smaller one was bifurcated into officers ward and offices. The operation room was built later in 1938. All barracks, inside have two rooms, one inner which is larger in size and other a small room, wooden planks and mud has been used in original construction and there is no trace of iron. The bricks were used later either in expansion or in restoration.

Fort from inside is all green, chinar, apple, pear trees are almost everywhere yet the four chinar trees at the four corners of the office barrack are worth appreciating; they were planted with precision and then looked after for long from grazing horses and other animals to reach such heights. They are as old as the fort itself. Fort has its own water supply.

Drosh Mess

Words at times loose their value because they are use daily and for everything thus the word impressive may look odd but this is what it feels to be sitting here at midnight when outside the yellow glow of full moon is getting pale .

A classic mess by design and style it is almost a replica of Petako Gaz mess in the colour and design. One small corridor then ante room on left and dinning hall on right with kitchen adjacent to dinning hall and billiard room/library adjacent to ante room. It is all in wood, which makes it so special. Good silver and trophies everywhere reminding of Lieutenant Colonel Murad Khan, It is much peaceful than Chitral, no more noise of motorcycles or busses only the constant melodic rhythm of water fall, plenty of greenery and flowers.

The hall is covered with walnut wood almost touching the ceiling all around, with hand carved fire places, shields on wall; PMA, Engineers, Punjab Regiment, FWO, head trophies of Markhoor stare from all corners with the head mascot on the southern wall keeping an eye on diner. Three silver trophies, the bronze infantry man holding and charging with the bayonet rifle, the big silver bowl in the centre and traditional ibex trophy without which no room of Chitral Scouts seems to be complete. On the far end at the fire place half a dozen silver ware and a tray below the Markhoor trophy, windows are on the southern side with ample space in the front to have the breakfast while having a look at the lawn and valley down below especially the Chitral Pass. Ceiling is also made of wood panels with chandelier hanging down , two ceiling fans and two lights on each end completes the scenery.

Mirkhani Fort - 1940

The fort at Nagur which was built by the local royalty in 1929 is impressive and looks like a military fort, it has one hanging bridge which connects it with the main road, at night the bridge is closed down; there is a post of scouts also to ensure this. The reason is obvious the other bank of river {southern} connects with Afghanistan and Kafiristan and as such it is closed down. In old days Nagur was one of the key places for crossing the river.

The Mirkani Fort is perched high on a ridge at the junction of tracks. Road from Drosh leading to Lawari is joined by a track which leads to Dommel Nisar onwards to Afghanistan. Thus Mirkani has strategic position in relation to the defence of Drosh and Chitral. No more wooden barracks or huts, only two exits one at the front and other at rear with another one on eastern wall which looks like make shift; a single storied barrack on both ends running all along. The length of fort is not more than 100 yards with a width of another fifty yards. *'I was wondering about the officers' accommodation when the sentry took my luggage and I followed him, we exit from the western gate and all of a sudden the full beauty of the fort was revealed. The area between the fort wall and the ridge which originally was meant for stables have now been converted into a beautiful garden having accommodation for troops and in 2009 a beautiful three room hut was constructed'.*

Being on higher ground the view is extended , with river almost 500 feet down below making its two turns, across the bank of the river the green fields of the Nagur with few house, Nagur Fort is not visible from here. However a look down reveals the sandy beach of the river which is presently used for playing cricket.

Originally the fort was meant to accommodate only one company and it did not have any mosque inside the compound. Presently it is hosting a wing, the mosque was constructed adjacent to the southern wall of the fort but after 2009 it was declared out of bounds for civilian population. For the civilian population a new mosque is under construction. The meteorological survey office is also located here alongwith the Pakistan Custom Post for Afghanistan.

Dommel Nisar Fort - 1942.

The fort was constructed in 1942, it is bigger than Mirkani but much smaller than Drosh. It is double storied, cemented brick, thick twenty odd feet high walls with high iron gate on one end and a small exit at the other end. It is 200 yards by 50 yards in a straight plain rocky ground with River Chitral running along its western wall separated by a narrow stony track. The fort is constructed in an elongated pattern mainly due to the dictate of the ground thus you end up facing the rather gradual high mountain wall. On the southern end the River Chitral takes its rather second last bend and merges to the west. On the east is {that is the back of the fort} high hillock blocks the view and provides protection.

Dommel Nisar is in the centre between the Mirkani & Arandu , it is the last fort built by British on River Chitral axis ahead are the posts like Langurbat and then Arandu, on the northern side coming from Mirkani the road is excellent but alas only for a brief distance, Frontier Works Organisation is working on the widening of the road till Arandu with United Arab Emirates sponsorship which certainly will bring a strategic change both economically and culturally. With the construction of road coupled with the opening of Lowari Tunnel the centuries old issue will be resolved, a loop will be provided to the Afghans and people living in Central Asia to move freely southward through Arandu Pass via Dommel Nisar-Mirkhani, and moving through Lowari in winter. It is the shortest, safest and most economical route. Other traditional route is through Zebak- following Bashgol Valley into former Kafiristan and present Nuristan of Afghanistan, hitting Arandu at Birkot. The earlier caravans had no option but to follow the River Kunar onwards to Jalalabad or the Kabul- Kunar conflux which is roughly 100 miles south of Arandu. The sectarian vermin, the majority of population living near Oxus is the follower of shia and Ismaili sect which the sunni tribes of Nuristan, Kunar, Jalalabad, Kabul, Bajaur, Khar, and Dir are not very friendly; much of present day violence in the area is mainly and chiefly attributed to this madness along with historical variations.

Dommel Nisar remained a quiet, isolated, lonely place where time seems to have stood still since it was constructed, it was awakened after the Russian invasion of Afghanistan {1979}; thousands of Afghan refugees entered into Chitral through this one route as all others were desolated. These Afghans made mud villages along the banks where ever it was available, those were the days that world was in favour of them and the people of Chitral welcomed them with open heart after all they were brothers; mostly. In 1962 Pakistan had exchanged the territory with Afghanistan , gaining the General Ayub ridge where a post bears his name across a mile in Torkham and in return giving them the area in Dost Muhammad which is across Arandu. Thus with this act the few hundred odd families all of a sudden found themselves as part of another passport, currency and laws. As a convention these people all along the Durand Line are given right to move freely across the border. At Dost Muhammad the situation is no different, there are 40 odd men of Arandu who daily go across the border to open their shops in Afghanistan by crossing over the bridge, reciprocally there are average 20 odd families visiting Pakistan daily just for medical purpose.

In the evenings the volleyball game is one of the most looked after event of the day. After sunset the Fort's lone gate is closed and life begins inside the fort, the mosque is outside the fort and it is only the Isha prayers which warrants the gate open. The additional battery of mortars and guns have extended the fort southward with new mud boundary wall coming up.

Posts & Piquets

Concept of post and piquet at Chitral is different from other corps of Frontier. Initially the posts were constructed for the Chitral Levies at Ziarat, Mirkani and Arandu and it was only after 1919 Third Afghan War that posts at Langurbat and Ursun were established. Russian invasion of 1979 made Chitral a hotspot. Arandu and Drosh were two too often hit towns. In 1986 landing of two Russian gunship helicopters at far north-west is one example of the diversified warfare of that period. These posts now acts as the cultural hub of the civilization in the remote areas, they provide medical and educational facilities.

Yarkhun Valley Posts

Kuragh Post. It was established in 1972 to low down the long distance between the Chitral Town and Mastuj,a staging point. It is situated 30 kilometers short of Mastuj, the site of famous ambush in 1895 Chhitral relief Force. Post is situated at the junction of three tehsils, Mastuj, Mulkhow, and Torkhow.**Parwak Post** was established in 1992 at the village of same name, mainly for temporary storage of items & ration received from Chitral on heavy vehicles and further distribution by light vehicles. **Mastuj Post** is 107 kilometers away from Chitral and is established in the village of Mastuj towards Boroghul Pass.It was established in November 1947. Post lies on the left bank of the Yarkhun River at an elevation of 7800 feet; which from this point onwards is called Mastuj River. 'Mastuj is properly speaking not part of Chitral, it has often been conquered by the Chital'. In the past Mastuj was goverened by relatives of Mehtar and known as 'Mehtarjao' who were independent of him. Winter are severe with cold winds blowing around the clock. An inscription at Barenis a neighbouring village across the bank of river reminds of past heritage when this area was part of Kabul Kingdom in 900 AD. Mastuj remained headquarters of the almost all wings of Chitral Scouts whenever they were posted here. **Brep Post** is situated almost 25 Kilometers north of Mastuj Post, primarily meant for security measurement and staging post. **Powar Post** established in 1979 is located at the centre of Yarkhun Valley almost 45 Kilometers from Mastuj towards Lasht Post. **Dubargar Post** is located ahead of Powar Post towards the Lasht, 70 Kilometers away from Mastuj in a village Dabargar. It was established in July 2001. **Lasht Post** is ninety kilometers away from Mastuj towards Broghul Pass, situated in the Village Yarkhun Lasht. It was established in July 1978. **Kishmanja Post** is located between Lasht and Broghul Pass post, usually house a section strength. **Broghul Pass Post** is 145 Kilometers away from Mastuj it was a front line post during Russian invasion of Afghanistan in 1979. It is in true sense a doorway to Chitral, winter are severe and physical links are severed, only wireless communication connects the post with headquarters. **Shandur Pass Post** was established in 1986 at Shandur the border between Chitral & Gilgit. It is 50 kilometers away from Mastuj and presently a jeep service ply between Mastuj and Gilgit during summer, height of the pass is 12500 feet.

Mirkhani Fort

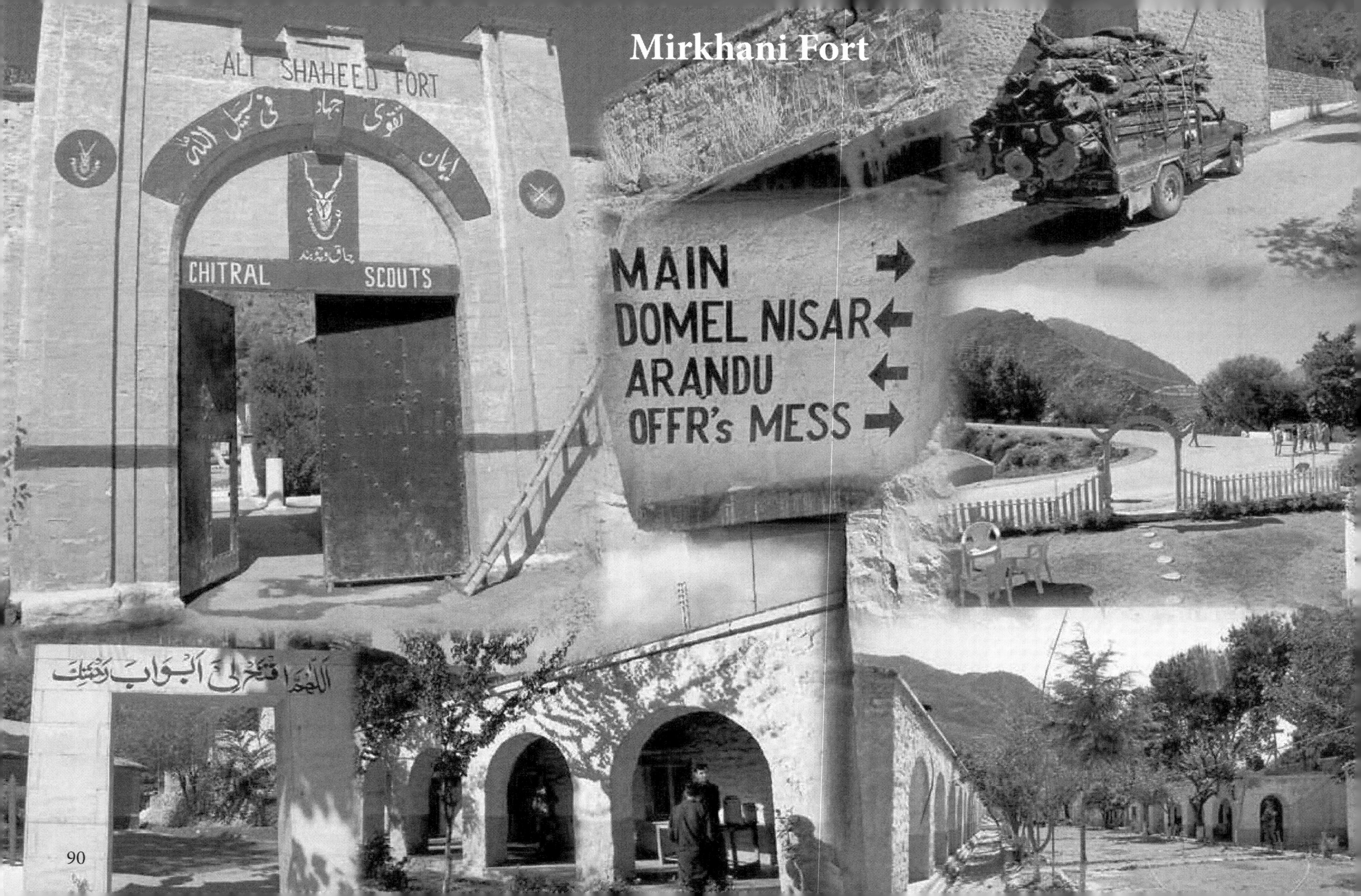

N
N/PORTAL
2700 Mils 12 Km
CHITRAL
200 Mils 50 Km
FAITHAZOM
BUMBORET
5900 Mils 50 Km
DROSH
550 Mils 11 Km
DAMMEL
3700 Mils 11 Km
LANGOR BAT
3780 Mils 19 Km
(GAMBIR) OP 26
3900 Mils 14 Km
URSUN
5580 Mils 4 Km
ARANDU
3850 Mils 24 Km

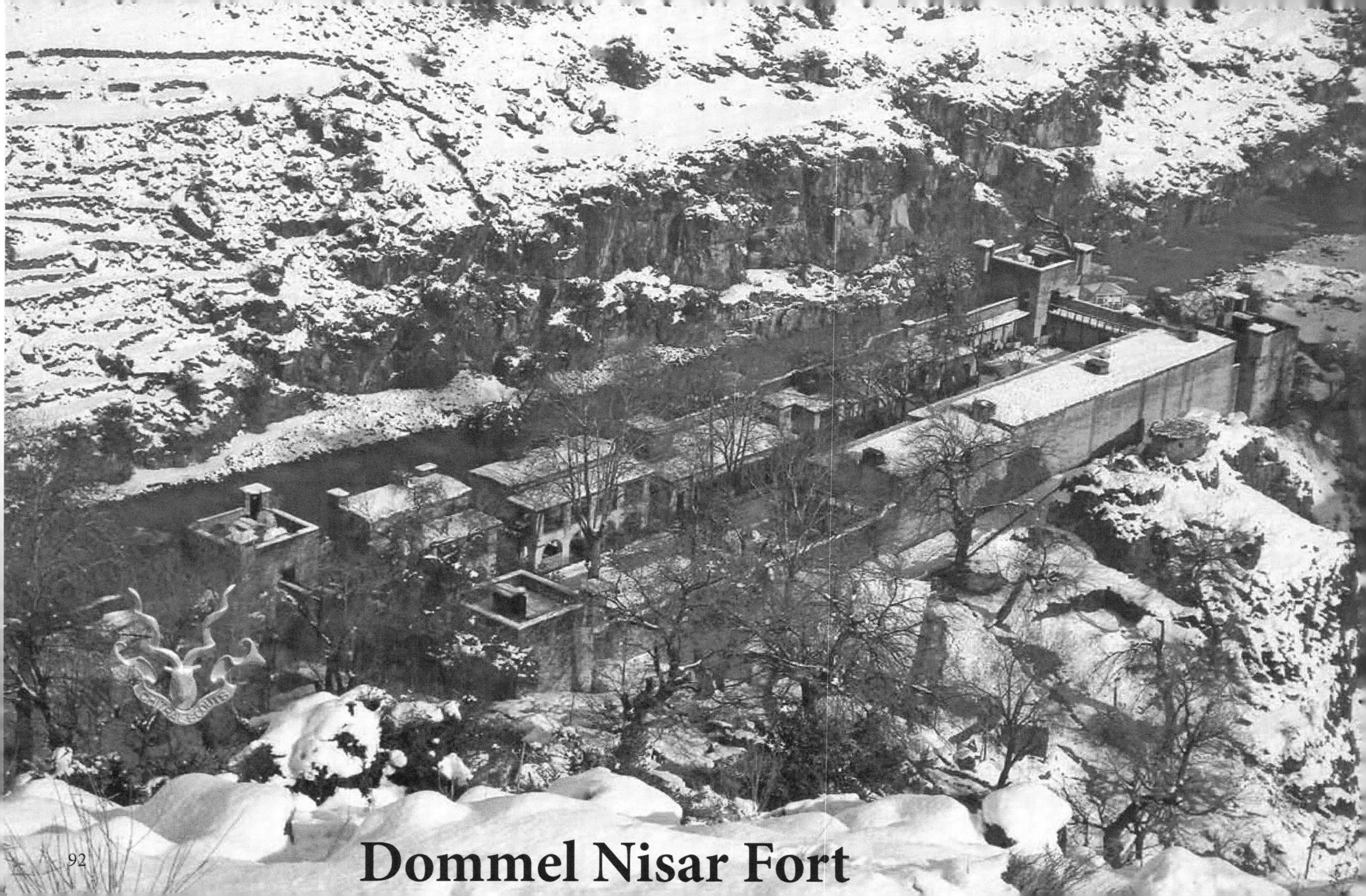

Dommel Nisar Fort

Dommel Nisar 1941

There is practically no rain in Yarkhun Valley, snowfalls fairly and lies for longer in Laspur Valley than else where. At Mastuj snow seldom lies for more than few days despite its elevation , due to strong winds blowing being at the junction of Yarkhun and Laspur Valley. Passes are generally close from last week of October with heavy snow fall between mid February to mid April; Broghul and Shandur are free from snow in mid June

Wakhi or Wukh- North Eastern Chitral

They occupy the north eastern part of Chitral in Broghul area, population is not more than three thousand, which are migrants who entered Chitral from Broghul Pass. One clan in known as Wakh or Wukh they are Afghans in nature who moved over from Wakhan area. Second clan is known as Sur Quli, these are Tajik and then those coming from Chinese Turkestan {Sinkiang} are known as Ghriz, collectively all are known as Wakhi and they speak a dialect which is different from Chitrali. Average elevation of the area is in excess of 7000 feet with pass towering at 11000 feet, the Wakhis are almost 200 kilometres north of Mastuj , their life is as primitive as it was thousand years ago, people still use horse, mule and donkey for travelling, there is no road or school other than one run by Chitral Scouts wing. Due to high elevation and extreme temperature there is hardly any cultivation here and people depends upon the animals for their survival, they prefer to shop or barter trade their necessities of life from Gilgit region.

Tirch Mir Valley Posts. (Injigan or Lotkow)

Shah Salim Post was established after 9/11 in order to secure the border, it is the main entry point along Dorah Pass into Chitral. The post is closed down during extreme winter. **Imeerdin Post** was established in 1981 during the Russian occupation of Afghanistan. **Garm Chashma Post** was originally established in 1987 to act as base camp for Shah Salim & Imeerdin Posts, now the wing headquarters is also located here. It is the site of famous hot springs water as well, thus in winter the troops have healthy environments. The track leading back to Chitral also closes down in winter. The place itself was a major tourist attraction a couple of years ago. Injigan is the western most valley of district Chitral, its literal meaning in local dialect is 'prosperous'. On its east is Tehsil Mulkhow, on west Afghanistan, in north the Hindukush and in south is Chitral Town and tehsil. TirchMir the highest peak of the Hindukush is situated here, near Karimabad, Rokhon is the second highest peak situated near Village Sainak, people calls TirchMir as the 'abode of fairies'. The cultural history and heritage of the Injigan is heavily inspired by the Tajikistan, Badkhashan, Yarkhun, Kashgir which is a logical conclusion of the valley being the entrance into the Chitral. The 'Jinan China' utensils are a speciality of the area which used to come from the Kashgir and China; they were almost unbreakable and above all had the uniqueness to point out the poison in food thus these were very popular with the ruling class. Silver Mushraba also comes from Faizabad in Afghanistan and are given as dowry utensils. Kohkan Baigi is the name of extra large big deg which were imported in thousands by the Prince Kohkan Baig and hence it carries his name even today. Ghaan is another utensil which is made of walnut wood it is big in size the smaller version is called Langri. Tong is a vessel used for keeping milk, lassi and curd, it is the smallest of all utensils. Ghori is the equivalent of Madaani which is used for making curd out of milk. Incidentally the capital of Fergana is also known as Ingigan. Manji or Manjani they live in Garm-Chashma in Lotkow Valley in villages like Para Beg, Postki and Sipohat etc. They also have Badkshani bloodline as they migrated from a place in Badhshan {Manjan & Yamgan} they speak Yadkha language which is an old Persian dialect and these people have preserved it despite adopting the local culture in totality.

Injigan is divided into three valleys namely Karimaabd, Urkari & Garmchasma. Weather is extreme in winter which blocks almost all ground routes due to heavy snow, spring starts from April onwards. Lotkow is the other name for Injigan, River Lotkow is the largest river of the valley which originates from Kotal Doawra and joins River Chitral near Chitral airport. River Bagosht, Othrai, Gol, Urkari and Karimabad stream all joins in River Lotkow before it terminates. The river is famous for its clear water and Trout Fish. There is mainly one crop in Injigan except in Bahtooli, Shigoor, Momi, Mough and Mardan Koh. The best grapes are produced in area Mough which are distilled to make local wine. Injigan has many other peculiarities but it is the presence of natural sulphuric warm water which is available year around; it is this which makes the Injigan as the most neat and clean valley in whole of northern areas especially when in other areas people seldom takes bath in winter. Mough is famous for the Chitrali patti the hand woven cloth. Urkari valley is the most under developed area of the Injigan, there was not even a jeep able track till 1983. Potatoes of Injigan are also very famous and in great demand. ***The very first person to be enrolled in Chitral Scouts in 1903 was from Yoft, Shaib Ali Lal son of Khokhan Bai***g, he later rose to the rank of subedar major the very first subedar major who was a non Kator[36] he retired, in 1932. Sultan Jawan alias Manoor, is another brave son of soil who took active part in the Gilgit War of 1947. Sultan Khan Murdaan was born in 1924 at Village Murdaan and joined Chitral Scouts in 1945 and took active part in 1948 War as Bren gunner, he embraced shahadat 40 kilometres short of Srinagar in a hand to hand fight with enemies. Another stalwart of the area and of 1947 War is Islam Shah who retired as naib subedar, he was enrolled in Chitral Scouts in 1932, he died a natural death in 1990. Ghair Dum Shah Charwelo who died in 2002 was another brave scout of Chitral. Injigan has a predominant Ismaili sect adherent, there are Fatimid also who are regarded as part of Shia sect but are not part of Ismailis.

Lowari- Arandu Valley Posts.

There are two major forts both constructed in 1940s at Mirkani and at Dommel Nisar. Ahead of Dommel Nisar is the post **Langurbat Post** and then the **Arandu Post** itself, it was constructed in 1942. These are old posts, Langurbat was reconstructed in 1979; due to present security environments few more posts have been constructed at **Darkot** and **Kauti.** Another old post is the **Ursoon Post** which protects the Kafirs from marauding Afghans. **Ziarat Post** is the oldest post of Chitral Scouts and it guarded and acted as search & rescue at Lowari Top. It had a well stocked library as well, with an instruction, not to remove the books. With the construction of Lowari Tunnel a new post has been established at the western mouth of Lowari{ eastern mouth is under Dir Scouts}. The **Lowari Top** itself is the boundary between the Chitral and Dir Scouts. A drive across Lowari is historic and a memorable journey indeed, there are no less than thirty two hairpins involved. The climb itself from the Dir direction {eastern} involves more ascend, scenery is beautiful with pine trees and fresh water flowing across the narrow track. In summer it is forded by the vehicles over the wooden planks placed by the natives over streams by paying nominal rent. The driving skills of truck drivers are worth appreciating over this pass. No movement is allowed after dark, in winter many have perished both scouts and civilians in crossing the Lowari on foot. Snowstorm, avalanches and loosing direction are the major cause of human loss. Lowari Tunnel at present is open for three days in winter and is closed in summer for repair & maintenance.

Arandu- Drosh {Gowari or Aranduni.}- Southern Chitral

The inhabitants of Arandu are a different class mainly due to lying on the gateway of caravan and traveller route thus the area is an amalgamation of centuries, a mixture of Afghan and Kafir along with the Mughul and Turk. They comprise of three small clans, first is known as Sinyari which

36 Muhammad Rahim, Tareekh Injigan, (Hafiz, Chitral, 2006), pp, 16-58.

Broghal

4985
3801
4358
Ochhili Pass
5553
Broghal
5560
Lasht
Kishmanja
6264
Garmush
4360
Shust
Darkot Pass
4590
5239
Dobargar
5884
Palut Ghari
Thui Pass
4575
Darband
Rech
Powar
Kotgaz Pass
5468
5132
Bang
Khot An
Khot
Lunku
Shagram
Brep
Zundrangram
Rayeen
Istoro Nal
7389
Shagrom
Parwak
5671
Zagar Pass
6431
Mastuj
4468
Sonoghur
Buni
Chumar Khan
Pass
4358
7690
Kosht
Kurag
Gohkir
Ghotbar Pass
5289
Reshun
Harchin
Susum
Beshgram
Baranis
Shandur

Lowari Top & Tunnel

The most important project in the history of Chitral, it allows Chitral to remain connected with Pakistan during winter also

Chitral Scouts

Lowari Tunnel

Chitral Scouts

Lowari Top

had roots in Asmar and believed to have migrated here almost five hundred years ago. Sultana is the second group which have roots in Jalalabad and probably came here almost eight generations ago, third clan is known as Afghani or Swati these have roots in Swat and adjoining areas, one of them lives here at Arandu and other at Dommel Nisar who are known as Damiri. The Damiris are located almost 30 miles north of Arandu valley, they constitute almost 1% of Chitral population. The clan has two sub clans, one known as Shuntri who claims to be the original inhabitants of the valley the second clan is commonly known as mere Swati or Afghani, they are nomadic in nature and have different residences for summer and winter.

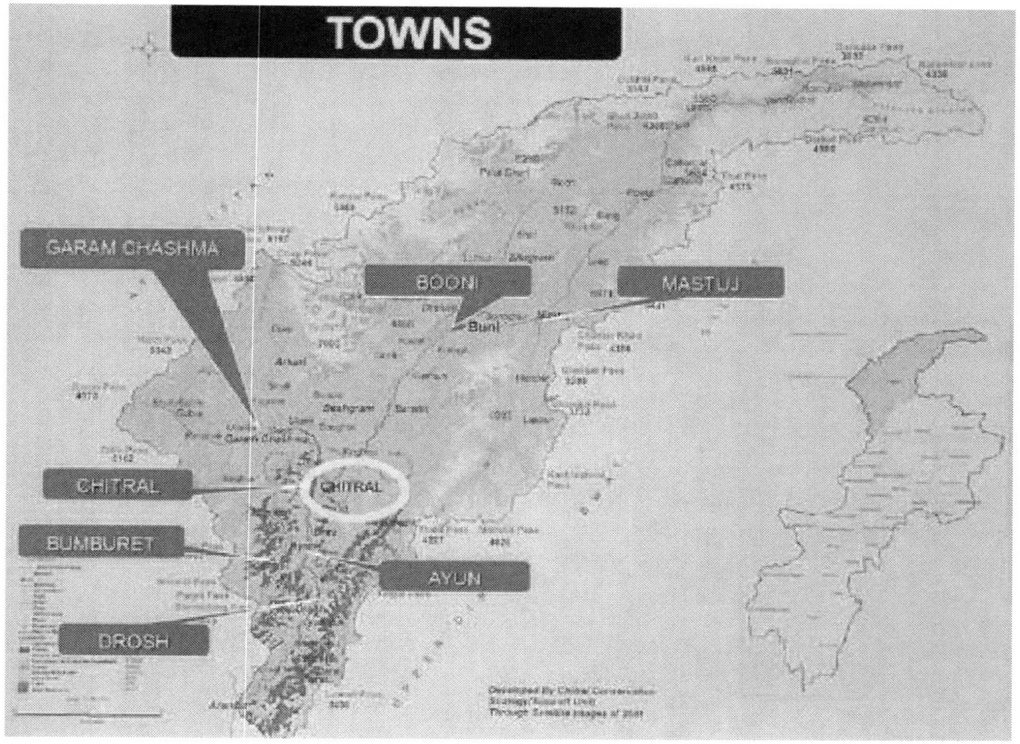

They all share same language which is different from Chitrali and share same cultural values. The highest crime ratio in terms of murder and other crimes occur here, mainly the revenge and blood vendetta. Their living habits are different the construction and design of their houses are tailor made for the blood feuds; high walls, narrow entrance and small windows, minimal education institutions are located here.

Tangiri or Dangkurk. They live in south of Chitral in the valleys of Ishtar Pewari, comprising of almost 1.5% population of Chitral, they speak a different dialect known as Phulwar which is known as Dungkurk in local terminology. They belong to the mountainous clans of Gilgit and surrounding areas and as such share the similar mountainous cultural similarities.**Gujar.** They are literally the Cow herdsmen, they are nomadic in nature and move with their herds depending upon the weather. Mostly residing in Swat, Dir and Kohistan area, majority living in Shisi Koh area north east of Drosh, their language is similar to Punjabi , they are quite clever in their dealings, mostly uneducated.

Pathan. These are mostly living in south of Chitral, some of them are also in Mastuj area and in Chitral city as well. The trade is in their hand, they speak their own language and generally remain aloof from the natives.

Kafiristan- Kalash & Bashgol- SouthWest Chitral

The Kafirs of Hindukush are the only living evidence of Chitral's past. The 3000 odd Kafir living on the southern bank of River Chitral in three thickly vegetated valleys with pine, walnut, oak, maple trees; the entrance to these valleys is through a narrow gorge, are an enigma. Tribes still following pagan rites despite being so close to civilisation. They are not man eaters and neither roam in loincloth, detest violence and generally remains merry. It seems as if they have been living in such remote, inaccessible valley since the time immortal. They at one time between 1000-1500 AD were powerful rulers of Kafiristan {Chitral} the very name Kafiristan was given by the Arabs to the area east of Badakshan{present day Chitral, & Nuristan province of Afghanistan} . The Kafirs had two main classification, the Red Kafir lived in the more fertile Western Kafiristan - Bash Gol{ gol is the common name for a stream, the valley takes its name from the stream which takes origin from Shah Salim Pass in north} and Kalash the Black Kafir or Sia Posh occupied the Eastern Kafiristan. They established the first Aryan kingdom, the passage of Aryans moving down from the high north took place through this geographical corridor{Bash Gol-Chitral- Kunar}. The Red Kafir were the ruling class the true Aryans and black kafirs were the result of intermarriages and working class. In 13th Century a Kalash ruler Bulesingha was defeated and driven away from upper Chitral by the Rais , a century later Rajawai the last of Kalash ruler was defeated by the Muslims{ Mehtar of Chitral}. Kafiristan retained its identity and Emperor Babur acknowledges it in his 1529 AD autobiography, Emperor Babur was aware of the presence of Kafirs when he first occupied Kabul. Rulers of Chitral ' have been in the habit of capturing their shia and Kafir subjects as well as Sia Posh Kafir....and selling them into slavery'[37]. In case of an offense the whole family was sold into slavery which formed one of the principal form of revenue. Chitral was even paying the annual tribute to Badakshan in terms of human slaves. Kafir slaves on escape were allowed to revert to their old faith by the Kafir elders. The price of a slave in 1867 was 100-500 rupees. World first came to know about them in 1885- 89 when first the British military commission reconnoitred the valley and later Major Surgeon George Robertson spent almost two years among these Kafir and wrote about them . Robertson thus stands as the pioneer explorer of Chitral , however he stayed with the Bash Gol area Kafirs who are termed as Red Kafirs. Durand Line affected the Kafirs most and more than any other tribe. Being non violent in nature they only had the geography as their best defence and now this was demarcated. Not that it matters on ground but in a broad term it took away the support of Mehtar from them as Chitral under Durand Pact was obliged not to interfere in internal Afghan matters and vice versa.

In 1895 the Afghans carried out a Jihad against the Kafirs, either accept Islam, fight or pay tribute. Kafirs in the past have been a popular commodity as slaves for the Afghans. Kafir's heartland is Bash Gol valley; the valley between River Chitral on north and Bash Gol in south, it is only at Chitral that area is bit plain. Resultantly the surviving Kafirs now inhabit the three ancient valley, Bhamboret, Birir and Rumbur however in the end they were almost exterminated in 1895 but managed to hold on to a narrow strip of land in lower Chitral. Majority of these Kafir embraced Islam and are known as sheikhs in society. ' *Some twelve miles south of Chitral city and west of Chitral River, there lies the entry to the gorge , between the high and rocky cliffs, the gorge slopes gently upwards, fanning out into a number of smaller valleys and it is in these valleys that the last surviving Kafirs dwell'*, commented by Chitral Scouts commandant, Lieutenant Colonel Afzal in 1972. Afzal was privileged to witness the culture so close, last one to do so was Major Robertson in 1889; Afzal has highlighted the Kalash Kafirs whereas Robertson had highlighted Red Kafirs. There are three main valleys, Rumbir which is closest to Chitral in the south , Bambouret is adjacent to it and Birir further south almost opposite Drosh. All are on the right bank of River Chitral and in a crescent manner occupies all area south, of Chitral.

37 Munphool Pundit, *Chitral, Gilgit & Kashmir*, 1869, p-36. Also see G.T. Vigne, *A Personal Narrative of a visit to Ghuzni, Kabul, and Afghanistan*, first printed 1842, reprinted 1982, Sang-e-Meel, Lahore, p-180, p-385.

Kafiristan

CHITRAL SCOUTS

Chilumjusht Festival

'Kafir tribe is a highly mixed people, I was struck on more than one occasion at finding a village where tall men with fair hair or light brown hair with pale blue eyes and a shorter type with black hair and olive skin existed side by side.... members of better classes showed signs of Aryans breeding in their good features ...a village chief in Bhumbirit told me 'our elders told me that our ancestors came from Iran and Greece....An elderly farmer in Birir valley told me that their ancestors came from Siam'[38].

Very little is known about Kafir religion and rituals. ' *Their original faith is a rather low form of idolatry, admixture of fire and ancestor worship. Imra is the supreme creator supported by a host of lesser deities; Giriz, Moni, Bagiz, Dizma, Krumai, Nirmali and others. Giz was the principal hero a virile warrior god. Kafirs believe in supernatural fairies known as Deo-Log, evil spirits is known as Shitan...one has to go high in mountains to invoke good fairies, the evil lives everywhere. Kafirs believe in hell and heaven known as baishat and dozegh respectively. Two particular divinities Mahandeo and goddesses Jestak { six feet high and eighteen inches broad wooden black plank} are held high in Bhumbirit valley . Fundamental ritual act in Kafir worship is the sacrifices of domestic animals, a proper sacrifice requires properly lit fire at altar, few branches of Juniper are thrown into flames with occasion twigs of holly oak, walnut or almond are also used; meat is distributed among the worshipperrs'.*

Marriage another social indicator of humanity is different too; every village had a male bull responsible for continuity of race. This concept is much closer to the Spartans concept of ideal warriors however here the intention and application seems to be different in the absence of any military laurels. Marriage an ancient social custom is celebrated in temple 'Jestak- Aan' by Kafirs. Bride and bridegroom enters the temple led by a goat, the goat is alter sacrificed by a boy virgin 'Jesta Mosh' , who later sprinkles some blood of animal on the Jestak as well. Bride has to eat five bread cakes before they are declared as husband and wife. Death rites are also conducted at Jestak-Aan, corpse is placed in a rough wooden coffin with a loose lid. For two days and nights the friends and relatives of the deceased stay at temple , eating and drinking, performing ceremonial dances around coffin, singing and dancing, in case of female dead body there no dance but singing. The religious priest is the most important person in the village, known as Shamman. He is an orator and performer as he has to perform religious rites in a opera manner. He creates hypnotic sensations around himself. Kafirs do not bury their dead, the coffin is left in open with stone lids on top at ' Maahan da Jao' {place of many coffins}. Giz is a red Kafir feminine deity protector of home, family and the private life, pregnancy, birth, children, love, marriages, sickness. In the Black Kafir valleys Sajigor, Jatz, Prebal and Warin does the same task. Rich Kafirs do erect the effigies made in wood over the coffins of beloved one. It is no more in use as old sculptors have died and so has the craft also. Kafirs of Birir valley are more orthodox and follow the customs more religiously, 1n 1971 there were only 2000 Kafirs living in Birir valley alone. Kafirs of Rumbir and Bhumbirit valley are beginning to give up some of their ancient customs and traditions; they now dig graves for their dead. Orthodox still do not send their children to school and neither entertain tourist as house guest.

Kafir women still stroll the Chitral bazaar wearing colourful attire hardly anyone can point out a Kafir man as he wears no distinct dress item. Women young and old all dress alike, a long black woollen gown 'Sangachs' tied in the waist with a woollen sash either white or light grey. On the head is 'Kopesi' an ornamental cap , heavily decorated with several rows of cowrie shells, hair most elaborately dressed in long and thin plaits. Kafirs

38 *Chitral & Kafiristan*, Lieutenant Colonel Afzal, 1972, p-82. Also see G.T. Vigne, *A Personal Narrative of a visit to Ghuzni, Kabul, and Afghanistan*, first printed 1842, reprinted 1982, Sang-e-Meel, Lahore, pp-234-239.

do not eat poultry meat and considers it impure.[39] Kafirs are fond of music but use only drums of various size and types, mostly women dances and men occasionally joins in ' The dancing party consisted of of all age groups, young girls, withered old women and mere children...they formed up into groups of four or five, each girl with her left arm around the waist of other girl to her left and her right arm across the shoulders of her partner on the right..cross and circles in a series of complicated patterns, displaying skills and natural elegance, rhythm kept by a seemingly wordless song with monotonous sound , from time to time we could hear the sharp and hollow thump of a big drum'.[40] There are three types of dances among Kafirs irrespective of occasion{death or marriage} and are performed simultaneously one after the other. First is known as Dosha, second as Drazhailak and last one as Cha. Kafir is a female dominated tribe, women do not observe purdah, remarkable number of women have chestnut hair with good percentage having blue eyes, they are forbidden to enter the sanctuaries of god except on special occasions, even the normal religious rites have preference to men.. They are not allowed to keep comb in the house rather leave it under a stone near a water stream, they set their hair after three days. During menstruation period women are segregated from the community for five or six days in a special house ' Bashaleni', every fair size village have one. Uncooked food is left on a stone opposite the Bashaleni and one of the inmates comes out and collects it. If for any reason any women from outside has to go inside Bashaleni then she has to strip naked before entering it and on coming out has to take a complete bath . Three main festivals of Kalash are Jyoshi {Chilimjuich} in May and last for three days. Women and girls gather flower and decorate houses, sheep and goats are sacrificed, dancing and wine drinking goes in every house, milk is sprinkled over goddesses Jestak. Porh is celebrated in September; it is the most ancient rite, the village male bull who was left in spring to live on the highland with special diet comes down and breed with women not bearing child, it is now observed in Birir valley only. Chownas {Chittermas} the last festival is held from 21st -31st December every year. Fire is lighted at sunset , goat sacrificed to Jestak, In Kafir mythology the Beromine{great giant} who lives in Siam comes only once a year to Kafiristan, he is the eldest of all giants and helps in all kind of issues, in matter of death, happiness and crops; his altar is known as Malosh.

Shekhandeh , these are converts from Kafir religion into the folds of Islam willingly. They constitute almost 1.5% of Chitral population. They are settled in Bhumbirit, Birir upper valleys like Langurbat and Jinjerait. They are also immigrants from Afghanistan, previously they were known as Red Kafir and were living on the western Kafiristan but after the 1895 Jehad majority of them embraced Islam, even now whenever any Kafir accepts Islam he is known as sheikh. Family life is compact and a normal family comprises of 15 odd members including the three generations living together. Male enjoys the polygamy and women works in the field, in short their cultural and social life has not been affected much by change of religion. Afghanistan after 1893 had carried out an all out massacre of the Red Kafir or Broghul driving them into the west {Chitral's Kafiristan}. Majority of the Kafirs have accepted Islam and are known as sheikhs moreover the Kafiristan itself became Nuristan in Afghanistan.

Madaglasht or Badakhsi- Eastern Chitral

Madaglasht is a north-western valley of Chitral or more precisely that of Drosh. Its fame is in its being a Persian speaking 2000 odd families living in a remote area. They were also famous for being arms manufacturer of highest esteem in the past and now for weaving and knitting woollen sweaters and socks, they speak Darri or Farsi language, they still have the customs of Persian culture but largely they have amalgamated with the

39 *Chitral & Kafiristan*, Lieutenant Colonel Afzal, 1972, p-82.

40 Ibid. p-76

Khowar culture.

There are two methods of going there from Drosh, one is the public transport, other is to hire the public taxi mostly Toyota Binjo cars, the jeep fare is rupees 200 . There are no hotels at Madaglasht and any tourist intends staying night has to lodge himself with villagers. The track leading to Madaglasht starts just on the northern edge of Drosh, initially the track is good as compared to other mountainous tracks but after twenty odd minutes it starts ascending, narrow, stony and dangerous. The Sheeshu River flows all along, it has clear water which is coming from numerous brooks, fountains, springs that falls into it on regular interval. At times its bed is broad but never is it too narrow. There are villages on other side too, village is a wrong word rather few houses or lone house perched high on mountains amidst the forest.

Route to Madaglasht

The 'Madaglasht' word holds true for an area but among them one village also carry the same name. This valley is peaceful and as such no requirement of even having a scout post. The scenery short of Madaglasht not only differs from the other in terms of the tree but also in the pattern of their graves, almost all graves are covered with wooden coffin something very peculiar to the Kafirs. There are three shrines also one of the very first that one comes across in Chitral. Jeep journey finally terminates at Madaglasht, a small mountainous hamlet.

'We stopped on the home bank of a wooden bridge and came across Naib Subedar Ali Nawaz of the Chitral Scouts he is retired since 2007. I was mentally ready to spend the night in the village guest room on payment but Ali took me to his home. Ali's home is small at least from the direction through which we entered, a small lawn having fruit trees , apples and pears with few space reserved for growing up of kitchen vegetables, the river flowing next to it. Ali was busy in harvesting his field but now he left everything and just concentrated on being a host. Ali have six children, four daughters and two sons. His two elder daughters are studying science, one in class 12th and other in matric , the third one being in grade seventh. His daughter Riffat who is studying in matric came and shook my hand which is something very unusual in this part of world but then this is what Persian culture is all about.

They all speak Persian, the young ones just jumping around in a pleasant way, no crying or shouting or abusing. Ali's cousin a post master who has recently retired also joins us. I learnt that whole village comprises of Ismaili faith Muslims and there are few Sunnis as well. I also noticed free movement of womenfolk which is so common in Punjab but so rare in this part. Evening was now growing along, weather was bit cool{Madaglasht is at 6000 feet}. We moved inside into the hujra, a medium size room having carpet and cushions on floor , pictures of Ali Nawaz in frames in uniform adores the fire place, few religious posters but nothing unusual in any sense. The washroom very neat and clean. I inquired about the electricity and water arrangement in the village. There is one hydel power house which supplies the electricity to the village around the clock, it is less but still better than nothing. Water is again coming from fresh water spring. There are two schools for primary education and one middle school and one college all less one are being run by the Agha Khan community welfare organisation. The teachers are good and qualified. In summer there is one month holiday in July and two months off in winter, January-February. In winter there is heavy snow fall and all routes are practically blocked. The other end of Madaglasht is also blocked with just a foot track that leads to the Goolen a small village north of Chitral. Thus people can travel from here to Chitral but it is only used by foreigner trekkers; no more there due to security threats. Postmaster narrated me that they all are Ismaili, their forefathers came from Afghanistan but their Persian is different from them, they had opted for this land . Apart from their language there is nothing peculiar about them on the surface but inside this cultural variation and openness is the key to understand them.

Children of Madaglasht

Next day return journey took place in darkness; the last vehicle to leave the village is around after Fajr prayers. It was a dangerous journey under moon light, still dark and driving through the mountainous track which is rugged and zigzagged. We climbed up the steep track then down, another sharp turn and then another sharp descend and this continued for an hour before jeep started staggering and soon came to a halt, it has ran out of diesel. We had practically blocked the road or track and now there were five or six other vehicles behind us , our driver got the diesel from them {six litres} by sucking through their fuel tank and this is how I reached back to Drosh by 0900 hours with very sore back yet I was full of admiration for the Ali Nawaz'.

Chitral - Myth & Superstitions

Chitral is full of myths which mainly comprise of fairies , there are Muslim fairy and Kafir fairy, there is a deaf fairy and then there is Kahngi Fairy which is an ally of housewives against their enemy fairies{Kafir & Deaf fairies}. For males there is 'Gor' a jin who is neither male nor female, the feet are turned way around with fire streaks coming from fingers. The 'Boona Jin' comes out at night, his power is in his cap, anyone who gets his cap becomes his master. There are no less than half a dozen fairies which are part of Chitral's cultural life.

In terms of superstitions the Chitralis are no better than other people and centuries old superstitious are still in vogue. If any one comes across a fox while setting course for a journey than it will be successful however if a shikari comes across a rabbit than he is doomed. If the right ear is getting warm than rest assure someone is talking good of you and opposite if the left ear starts getting warm. Presently there are two much talk about personalities in Chitral, one is a female, known as Parri Begum who is believed to know the palmistry and can recall the souls of dead ones. Other is a pir in the high mountains of Mulkhow , who again has the power to mould the future; matrimonial affairs is his specialisation and his six wives are a living publicity of his craft, he charges 6000 rupees for a visit.

Kohistan is the name of area east of Chitral and Yagistan is another name for Dir and Swat area. Kohistani language is different from Pashtu and is more akin to Punjabi, they are also converted from Islam four or five generation ago{in 1860}. Dress is similar to Chitral and Gilgit having a long shirt and brimmed hat, their winter shoes are very peculiar and known as 'Pajola'. Kohistan women work with their husband in field, they wear woollen-silken head dress. Kohistani are migratory in nature in their folk lore they say that they have learned three things from Afghans, the use of leather shoes, use of long swords & guns and also the system of betrothal.[41]

Women

Culturally women enjoys a high esteem in the society , there are instances of women being appointed governor during the reign of Aman Ul Mulk, furthermore even Shuja ul Mulk was married to a widower. They follow own traditions in marriages etc, like most of Pathan tribes here also the groom has to pay for the marriage expenditures including the clothing of bride. There was a time when the Chitral was a hub of white flesh trade but it is no more there. Chitral is facing a dilemma of social values, the majority of men from lower areas prefer to get married here , since 1995 more than 350 cases of marriage outside Chitral has been taken into the courts by the local organisation in which it was revealed that only 270 marriages have taken place within the family rest all were outside the family in which the groom had paid the amount and move away with the bride. The important thing to understand is that Chitrali women are beautiful, cultured , well mannered, obedient, naive, innocent and family oriented and as such are held in high respect as a potential wife and mother and this point is often misunderstood by non residents. Chitral girls and boys were the most prized of all the different descriptions of slaves brought to Turkistan market excepting perhaps Persians for their beauty. The very first girl's school was opened in Chitral in 1936 known as Anglo Vernacular high school.

---

41 Dr G.W. Leitner *The Races and Languages of The Hindu Kush, Imperial and Asiatic Review. July-October 1891.* pp.139-156.

Culture & Weapons

In terms of weapons a sword is called a Khonghar, it is very thin at the edge and in old times it was always dipped in poisonous water before the battle. Shield is known as Khaari, Bow & Arrow are called Dhoorwan – Weishu,Nairang is a battle dagger worn around the body. Dumpla was a firearm using cartridge and Dumpla por and Rupal are rather the basic forms of firearms.The weighing measurements are also different, Batti is 2½ kilogram, Kondak is ten kilogram, Beera is 20 kilogram, Wara is 60 kilograms. Aichamoot is smaller than an inch, Dasheet is smaller than a foot and Hoost is larger than a foot but smaller than two feet.

In terms of dress the Injigan is famous for woollen cloth , caps for male and female are both made here, female cap is colourful and a must item for the bride known as' khoi suak'. The woollen floor mat is known as 'Pelsibick Koreek'. Chitrali months are related to the culture for instance February is knwon as Urian for the birds {duck}that arrive here, name of days are as in Persian , Shaanba, Do shaanba ...etc.

A polo player trotting through Chitral Bazar

Polo & Music.

Polo in true sense is the identification of Chitral , it is played here in almost all village 'maidan' the way it was invented and played millenniums ago; no rules. Polo took birth in Central Asia with the domestication of this fine specimen of animal, the horse. These mountainous breed was and is still regarded as the best in the world in terms of stamina and intelligence. Key to success of Aryan was their mastery and love for horse, they initiated the very first Blitzkrieg in the history almost 1000 BC and kept rolling by Mongols and Turks most famous being Mahmud of Ghazna, Chengiz Khan, Halugu Khan, Tamerlane and Babar of Fergana. Game itself originated as a thrill & fun which comes up naturally when Aryans felt the speed, excitement and competitiveness while galloping on the steppes. Mongols particularly Tamerlane was fond of playing it with the skulls of the vanquished. Buzkashi is the only other game which can rival Polo fervor. It is another strange twist of history that both these games are played in traditional pattern {which is nothing other than living natural} only in Hindu Kush{Badakshan, Chitral & Gilgit}. Polo is not played in Badakshan and Buzkushi is unseen in Chitral.

During summer which is short but warm , at evening one comes across Polo players trotting through Shahi Bazar towards one of the three Polo Grounds in Chitral, same scene can be seen in almost all villages of upper Chitral. Grounds are uneven, barely wide and long enough in high mountainous altitude of Chitral to allow any kind of safety to the horses, players or spectators. It is like an arena, small few feet high stony walls with sheer falls, defiles, gushing water from gols, back ground of mountain, centuries old maple or walnut trees adding to the scenery. There is no restriction on number of players, game duration lasts for 45 minutes to one hour without any change or rest for horse or player. There is no protective headgear or clothing, you just sit on the horse and gallops after the ball, hit it as far as as you can{ that is if other team players allow you to come even near the ball}. Pushing the horse, blocking the shot, cutting the line is all fair. In fact Polo is the most Gentlemen game in true sense, the opponents test you but intention is never to hurt or harm the player or horse.

Music is another important segment of local life and especially no Polo game is complete without the presence of two instruments the drum and flute. They are soft in nature and provides perfect ambience to the occasion. Spectators sitting on the walls practically feel the heat of horses, their eyes remained glued to the ball for their own safety. When a player scores a goal then he picks the ball and strikes back towards other end, it has its own glamour. The player taking the shot after the goal does it with his own signature tune. The musicians knows it and when the sound travels in air it lets the mother, sister, wife and other beloved ones who have not been able to witness the game to know that he has scored a goal.

Shandur annual Polo match between Chitral & Gilgit is a classic, at 12000 feet high plateau which connects and act as the natural, historical and administrative boundaries of these two rival riparian mountainous cultures The River Gilgit & River Chitral {Yarkhun}. It started in 1903 and oral

Chitrali Dance

Chitral Scouts Musical Troupe

history traces a love string behind it, one of the princesses condition of marriage was 'the one who wins at Shandur' . another oral tradition traces its back ground to ancient times when the village democracy resorted to Polo for arbitration rather than the violence. { Chitral & Gilgit have one of the lowest crime rate in the region}. In one of the epic matches which natives still love to talk about was on 28th August 1966, final was played between the Chitral Scouts and Northern Scouts, it was won by the Gilgit team, who had earlier defeated the Chitral Scouts in the preliminary match as well 'Gilgitis were no doubt better than us in every department of the game and it is a fact that we have to learn a lot from them'[42]. In that year {1966} polo season galloped off on 12th October, Chitral Scouts entered two teams, first team was headed by commandant Lieutenant Colonel Shaukat Sultan which fought its way to finals but eventually lost to the all times winner Zagrazar, the deputy commissioner's team comprising of natives. The other polo team of Chitral Scouts lost in opening match to Balach another local team. This was Lieutenant Colonel Shaukat Sultan's last polo tournament as he was posted back to army in November 1966 after completion of his tenure, ' He was our commandant, infact a very special commandant, but it is as an elder brother that we miss him and shall remember him most'[43]. He was replaced with an equally special commandant, Lieutenant Colonel Muhammad Sher Khan, Sitara I Jurat. 1986 Polo match at Shandur was historic, President of Pakistan General Zia Ul Haq graced the occasion and Chitral Scouts commandant Lieutenant Colonel Murad Nayyar was the man behind this. Since then it has become the most prestigious event in world Polo calendar. It is played between Chitral Polo Team and Gilgit Polo Team. At Shandur Polo is played in most natural manner, duration is 45 minutes without any break, player cannot change his pony, a pony has to sustain the entire duration. If a player gets injured then other player can replace him but pony will not be change; on the other hand if pony himself becomes injured than the opponent team has to play with one less player. It is in fact a test of horse and rider both in stamina and endurance apart from intelligence. Chitral is not famous for horses, they all come from Badakshan in past and now from Punjab.; thus it is logical to conclude that where as Polo originates from Persian speaking Aryans the Buzkashi belongs to Turk. Within Chitral , Polo is played in upper Chitral only which is more Persian oriented culture, in lower Chitral, it is rarely seen. Kafirs have no tradition of playing Polo and neither it is played in the former Kafir valleys. Game is alive in Chitral, it is not dying neither its future is in danger. It is the pride of Chitral, Chitral Scouts and natives are keeping it alive; it provides them much needed adrenalin in a non violent manner.

General Zia ul Haq 2nd from Left at Shandur 1984

42 Frontier Corps Newsletter 1966, Chitral Scouts, page 31.
43 Ibid.

Chitral Scouts Polo Team 2012

Shandur, 12205 feet, is the highest Polo Ground in the World. In 1936 the very first Polo Tournament was held, today it is the most famous festival of Pakistan.

Headgear & Mascot

Chitrali head gear is very peculiar and is the identity of the area it is also part of Chitral Scouts ceremonial uniform. Every visiting army chief has worn it notably President General Zia Ul Haq, General Waheed Kakkar and President General Mushhaarff. Lady Diana has adorned it so has Prince Karim Agha Khan. It is part of uniform, now a day it is worn on ceremonial occasions. The cap itself is made of Chitrali cloth and the Village Mogh near Garrm Chasma is famous for its quality. Cap is custom made, taking the measurement of the head and then cap is sewn; it is available readymade as well. In Chitrali language it is called 'Pakool'. The plume is made of duck which is known as Mundhaqq in local language, it is a seasonal bird , however from 2002 onwards the use of natural feather has been prohibited keeping in view the endangered specie. The plume now is artificial, however the original plume is known as Putcch in Chitrali language.

The cap is white in colour, which is rolled up consuming more than a meter of cloth, presently it is not issued free of cost to the scouts, they have to purchase it from open market where it can cost 250-500 rupees. The Gilgit Scouts have identical headgear also the only difference is in the way of making the cap, Chitrali cap is more heavy and have more twinges as compare to the Gilgit Scouts ; which is bit narrow and its top cloth does not spill over the rim. Mahsuds wear the same kind of headgear with minor design and wearing pattern. A Pathan seldom wears this headgear he has its own white skull cap in summer. Mahsud style of putting on cap is distinct, they place it in middle of head {Ahmed Shah Mahsud 'Lion of Panjsher' made it famous}.

The Chitral Scouts mascot is Markhoor which is again an endangered specie now; It has distinctive cloth background which denotes the wing colours {presently there are six wings in Chitral Scouts}

Language.

All Chitralis are not same, their faces, colour of eyes are different and so are their customs and traditions, however the language and religion is the biggest binding and after it the Chitral Scouts bondage is the most powerful institution in the Chitral.

Princess of Wales Lady Diana wearing Chitrali Cap - 1992

Khowar is the language of Chitral River from snout till Mirkani in the south with varying dialect, in its pure form it is being spoken in the upper Chitral in Mulkhow and Torkhow area ; ironically it does not have any written characters thus Persian remained the official language since early days later replaced by Urdu and English. Khowar is like Urdu because it is also a combination of Persian, Turkish and Sanskrit language. Till mid seventies a journal in Khowar language was monthly published by the Government of Pakistan to promote the language. In Lotkow, Madaglasht and Kalashgum the dialect is different. Persian is still spoken and understood by the aristocracy in the Madaglasht in the Shishikuh Valley, Yudgah is spoken in upper Lotkow Valley above Parabek, Dangarik or Palola is spoken in Ashret. Kalkatak , Beroi and Nastiwar or Gawarbati is the language of Arandu. These varieties of language when seen in the context of the Waziristan which have much more area but only two dialect reinforces the theory that Chitral has been abode of many races and cultures which with the passage of time adopted a working pattern rather amicably. Arnyia is another name for language spoken by Chitralis

The abundance of fresh water and green pastures all along the various streams in Chitral provides a solid logic against the utter use of violence for the land as observe in the Waziristan.

Chitral Scouts Gardeners wearing Chitral Headgear

Hunting & Fishing

Chitral is famous for the hunting , fishing and bird shooting. It is a migratory station for the Siberian birds and as such are much awaited by all. The hunter and hunting have their own customs and traditions. In Chitral, men spent much money in the hunting season often falling into debt as well. The main cost is in the construction of a hunting piquet at the river and other lakes . Hunter then sits inside it at early morning, float the decoys and then it is a matter of patience before cranes starts landing. In summer almost everyone carry a fishing stick/rod and fish is also available in market. The protein requirement of natives is thus fulfilled through this activity. Big sport is highly restricted and permit is very expensive, snow leopards are quite a common sight.

Chitral Scouts and War against Militancy

CHITRAL SCOUTS

Chapter Six

Chitral Scouts and War against Militancy

Chitral Scouts is actively participating in the ongoing operations against terrorism since last three years. Chitral Scouts has kept its flag high in all operational areas i.e. Swat, Bajaur, Spina Thana (Darra Adam Khel), Dir Maidan and Orakzai Agency. Chitral remained safe from the flames of terrorism. It was in August 2011 when hell broke loose in lower Chitral when the physical attack was launched by the Afghan Bashgol . Chitral Scouts is organised into six wings, with a mountain artillery battery and mounted infantry as integral components. Overall command is with a colonel who acts as commandant. Chitral Scouts was the first corps to have a colonel as commandant. Each wing is commanded by a lieutenant colonel, having four companies each commanded by a subedar. Although these companies should be commanded by regular army officers but there is always a shortage of them. Furthermore each company comprise of four platoons each having three sections. Platoons are commanded by a havildar. Strength of a wing varies from 600-690 scouts. These wings rotate after three-four years among the various stations in Chitral like Drosh, Mirkhani, Dommel Nisar, Mastuj, Chitral and Garam Chashma. They manned posts which are outlying like Broghul Pass, practically they have posts on every pass and in every valley. Task of post vary for instance the posts along Drosh-Mirkhani-Dommel Nisar and Arandu have more concentration towards anti-smuggling role where as in Broghul and Shah Salim Pass they are more like listening post, these are in fact the symbol of state. These posts provide search and rescue in winter, disseminate education & medical to locals; overall peace and tranquillity seldom required any scout action in the area.

The pattern and conduct of war is unconventional, miscreants are small in number and mostly based in Waziristan, from there it slowly and gradually started affecting adjoining areas. These miscreants are not under any uniform command rather it is the last two hundred years of history being repeated again. Dir, Swat , Buner were the strongholds of Ahmed Shah Barelvi's followers {1825-1860} and they had nothing to do with the Wazirs. Violence in Kurram and Orakzai Agency{Orakzai Agency was carved out of Kurram in 1974} is mainly around sectarianism. Bajaur was also once part of Khyber Agency and violence has its roots in Afghanistan. But by and large the violence is restricted to Pashtu speaking areas and even more prominently in tribal agencies. Maulana Fazalullah of Swat is the one calling the shots in the area.

Violence within Chitral has always been restricted to the Arandu-Drosh axis with the exception of 1926 when Lotkow and upper Chitral were targeted by the Mehtar on religious grounds. Mostly it is the Afghan issue which is the root cause, the issue again is not at national or international level rather it is the local geography & history which is the firewood of the violence. 18th Century campaigns for conquest of land or the crown or even that of egos have almost exterminated in last fifty odd years. The Russian invasion of Afghanistan in 1977 was a big test but even then Chitral was a normal district. Russian used to bomb Arandu and Drosh but no physical attack was launched either by them or the Afghans or the Bashgol on their behalf.

In last twenty years the sectarianism has emerged as the biggest threat to the tranquillity of millenniums in Chitral. Sunni,Shia & Ismailis have been living here in such harmony which recently was matched only by Gilgit. Similarly not long ago, only few years ago the foreigners were roaming in the bazaars of Chitral, lone woman travelling at night with porters, scholars coming from around the world to explore the Kalash Valley; all this including the mega projects like Lowari tunnel and Goleen Hydel power project are now under threat of miscreants attacks.

Chitral Scouts artillery battery, mortar battery, signal platoon, mounted platoons have all been supporting the operations against the militancy. It is the beauty of Chitral Scouts that despite all the ongoing operations the life still maintains its normal pace. Schools have been upgraded, houses for shaheed have been constructed, constructing parks for children, conducting the annual Polo at Shandur{it was not held in 2013}, sports tournament, promotions of scouts, training for recruits, dine out of officers and subedars, expansion, extension, preservation of forts and old heritage. Free medical camps for the local and providing protection to all the major projects in Chitral. Protection, promotion and nurturing of Kalash people & culture is also part of Chitral Scouts.

Athanasius of Bhumbirit - September 2009

Kafirs of Kalash claims to be descendent of Greeks and it is very much possible also because Alexander The Great after his retreat had divided the India among two of his lieutenants who remained in power for another three hundred years . The way to Oxus from Indus or vice versa has to pass through the Kafiristan; not necessarily the Chitral River. Thus Greeks or Hellias as they call themselves now have developed a cultural heritage with them , it serves both as Greeks are also looking for something to revive its old glorious history and Kafirs looking for more attention. Greeks have developed the museum, water supply in the Bhumbirit which is the largest of Kafir village, community toilets and few other minor things. Mr Lerounid Athanasius was one such person who was living in the village since 2002. On 7/8 September 2009 at midnight, two men overpowered the Head Constable Zafar who was there as part of security to Athanasius; heading four men. It is worth mentioning that not before this, had ever any policeman stayed a night here because of the zero crime situation. They at times would close the police station at sunset, hand over the keys to the local headman and then come back in the morning after. Constable Zafar died when he hit the ground, other two policemen were injured and Mr Athanasius was taken away.

Chitral Scouts platoon under the command of Major Naseeb arrived at first light , all bridges and tracks leading to Afghanistan were sealed, border police was alerted and local jirga was called and addressed by Major Naseeb. Residents of Bhumbirit, Ayun, Urtsun & Birir serving in Chitral Scouts were sent on leave to gather the information from the villages.

The first lead came from Noor Muhammad a local who disclosed that he had provided mules for some Afghans

'I was going from my village Otak Banda to grazing ground at Waler Banda with my mules when I encountered a dozen Afghans along with a person wearing a chaddar moving under their protection they were climbing up, the man with chaddar was walking with a stick other were carrying weapons, one man told me rather ordered that he is their officer pointing towards the man wearing chaddar and I should take him upon on my mule ,I

refused and was taken along on gun point...after some time my animal also got tired and I was ordered to carry the officer on body which I did..later I really had to touch their feet to let me go which they did. It was Ramadan and they were eating and offered me the same but I declined. I had talk with them and asked about the person , they narrated that they have abducted it from Bhumbirit and in the process had killed one policeman also, they had wrapped the bombs around their body and I think even the bag which they were carrying was full of bombs'.

One cannot ascertain the truth of this but it was certain that they adopted the Parapit Pass route to reach Patti Gol. Commandant Colonel Rizwan Rafi { former special services group} visited the area and held the jirga and was of the opinion that local residents of Bhumbirit & Shekhandeh are involved{it proved correct later}. The parties of local were sent to Nuristan on 10th September comprising of local jirga members for release of Athanasius , one left on vehicle via Arandu-Birkot route the other two on foot via Zingoor Pass.

On 2nd October 2010, the jirga came back with a letter from Athanasius, which was delivered to the Chitral Scouts, in which Athanasius wrote 'I am fine here until today..Taliban demands 20,000 US$ and freedom of their prisoners'. Two of the messengers sent to the local Taliban leader had an unfortunate accident as one was killed when NATO aircraft bombed the area and other was seriously injured.

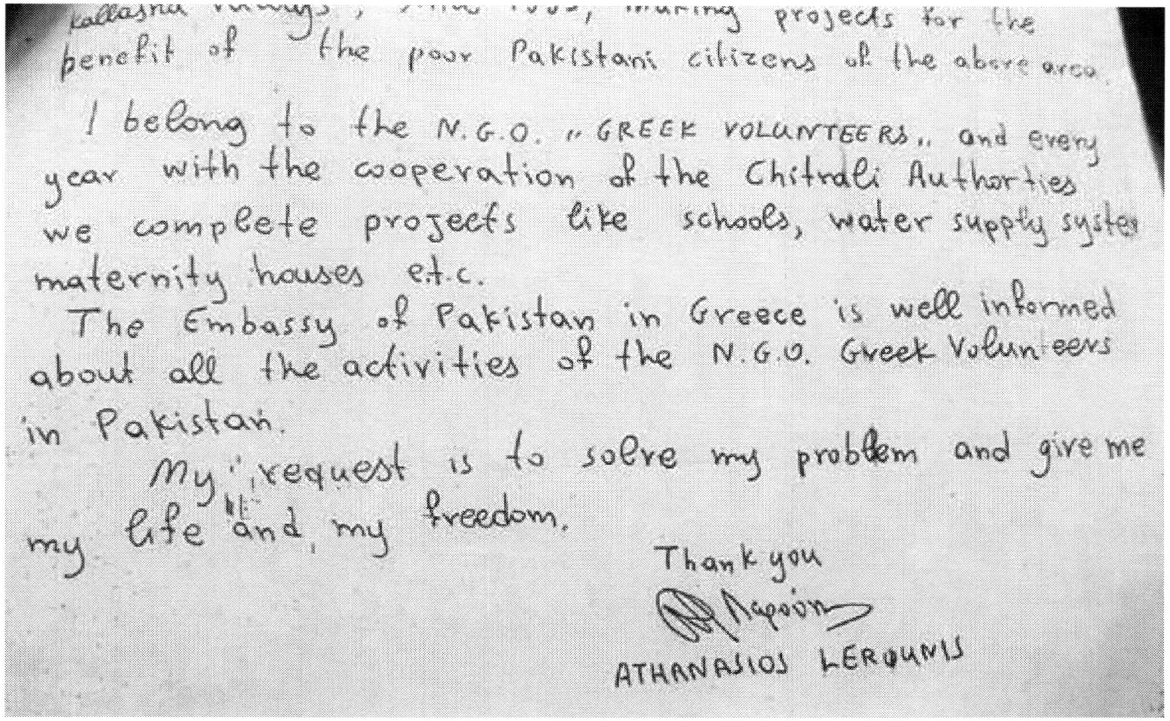
... projects for the benefit of the poor Pakistani citizens of the above area.

I belong to the N.G.O. "GREEK VOLUNTEERS" and every year with the cooperation of the Chitrali Authorties we complete projects like schools, water supply system maternity houses e.t.c.

The Embassy of Pakistan in Greece is well informed about all the activities of the N.G.O. Greek Volunteers in Pakistan.

My request is to solve my problem and give me my life and my freedom.

Thank you

ATHANASIOS LEROUNIS

Letter of Athanasius

Mr Athanasius was later released by the Taliban after Greek Embassy agreed to pay the ransom, however prisoners were not released.

Zhangshall Abduction. On 28th November 2010, eight men were abducted by the Afghans while they were cutting woods, they left one person with the message that Zhangshall belongs to them and no wood cutting is allowed . Later all the wood cutters were released who narrated that they were taken into Gawardesh in Afghanistan where the abductors simply got tired of feeding them and thus left them one by one. The woodcutters were hired by the locals for wood cutting and had come to the area on vehicles. Lower Chitral is famous for the wood thus a 'wood mafia' has emerged. This small incident highlights the intriguing face of the conflict. Everything which happens does not necessarily have roots to extremism.

CHITRAL SCOUTS

Recruits Training

Qasam Parade - 2012

Ursun Post - 1st May 2011

On the night of 1May 2011 the miscreants belonging to the Swat chapter attacked the post which was held by 30 Scouts of Chitral at 0200 hours, miscreants were in strength of 20-30 who had infiltrated through the Paith Zom Pass and Gambir Gri Pass to cross the frontier and then used Banjal to reach the post. The initial plan of the miscreants was to attack silently but they were discovered by the alert sentry and that resulted in exchange of heavy fire. Five rockets fired by them landed inside the post, this fire duel continued for another three hours before the miscreants were forced to flee. The miscreants while fleeing attacked one post of border police and took away eight SMGs along with three policemen of border police as hostage. Miscreants were believed to be part of Fazalullah Group[44].

Darashot- The Longest Day.

On 27th August 2011, the Afghan miscreants attacked seven different posts of Chitral Scouts in Arandu – Mirkani area spread over 32 Kilometres, the crow distance between them is 22 kilometres; soon after Fajr prayers. It was the first such attack of this magnitude since 1919 and when seen in present context of war against militancy this was the dragging of Chitral into the Durand Line affairs from which it has been able to keep a safe distance since 1895.

Darashot

The attack lasted till noon, posts which came under the attack were Ursoon, Mirkani, Dommel Nisar & Langurbat with heavy fire, mortars, rocket propelled grenade launchers , all which is usual in the tribal attacks on the posts all along the front and physical attack was launched upon the Gudigar,Darashot and Kauti. It is the pattern of the Taliban in which the attack is launched on a group of posts , keeping some under the fire and attacking one particular physically. It creates delaying action upon the defender especially at night in mountainous terrain with very poor communication tracks. Chitral Scouts stood out the test and fought bravely holding the posts intact especially at Darashot.

44 District Coordinating Officer Chitral, Letter No.4225 dated 1st October 2010.

People living in the area especially at Nuristan, Shekhandeh, Bhumbirit Rumbur, Ursoon, Ginjirate Kuh, Arandu, Chitral are relatives to each other since centuries and they have been travelling almost daily to meet each other and carry out the chores of centuries old rituals of trade through frequented and unfrequented routes[45]. The district administration had established the contacts with the people living in Nuristan for maintaining peace in the area and not allowing the Taliban or hard core Islamic militants in the area; so far it had worked for almost eleven years a remarkable achievement when seen in the context of overall devastation of social fibre in the frontier and other regions of Pakistan.

Miscreants had the concentration area at Gawardesh from where they crossed into the Chitral boundaries through Pathkun Bro, Paitazom, Kauti Gol, Gambiri Gri and Chulu Gol. They were a mixture of Afghani and Nuristani men wearing stolen NATO uniform and definitely supported by the Afghan National Army personnel as one of the dead body later revealed. It is highlighted that in May same year five posts of Afghans were wiped out by the miscreants which Afghan alleged to be based in Pakistan. The warning about the impending attacks was received almost a month ago with rumours coming from the lips of the travellers and the centuries old natives' way of information. What was surprising, was the magnitude as almost seven posts were hit simultaneously. The quick reaction forces were made ready at nodal points and wings were warned to keep a high alert and this in the end saved the day. The four worst hit posts were Darashot, Gudibar, Kauti & Langurbat.

At Darashot thirteen individuals of the post including post commander embraced shahadat, yet they did not surrender it , Sepoy Nasir manned the machine gun till 1300 hours when the relief force came on the post. Attack was led by the Ghazwan and Zubair both local miscreant commanders with Zubair getting injured also.; all in all 32 Taliban were killed , the remaining while fleeing burnt the hanging bridge on River Chitral thus practically bringing the chase to an end. They had looted the weapons as well which they took it back via Langurbat, Arandu, Khur and Dab.

Sepoy Nasir Min Ullah

Sepoy Nasir Min Ullah- Last Man Standing

On 27th August which was 27th of Ramadan also, Sepoy Nasir after taking his sehray along with six other scouts was reciting the Holy Koran. No joy can match the recitation of the

45 DCO Chitral, Letter No.2818/DCO/dated 6th September 2011.

Mayoon
Ursun
Mortor Position
Arandu Fort
Gambir
UMBURET
PASS
URTSUN
Domel
Arandu CP
Shaheen
Arandu Dildar

SHUHADA_E 2WING AT DOMEL_ARANDU SECTOR
(27 AUG 2011)
GUDIBAR POST
HAV BAJGI KHAN
NK NAZIR AHMED
SEP SHER AKBAR
SEP TARIQ MEHMOOD
SEP FAQIR HUSSAIN
SEP HAZRAT ULLAH
LANGURBAT POST
LNK REHMAT NIZAR KHAN
SEP ALI HAIDER
B/SEP AMIR ZAFFAR
KAUTI POST (1WING)
SUB YOUNUS ALI
DARASHOT POST
SUB NIAT GHAZI
HAV ALI MUHAMMAD
HAV JALAL UD DIN
NK MANZOOR ELLAHI
LNK SHAHZAR WALI KHAN
SEP ZIA UL MULK
SEP HASHIM PANAH
SEP MUHAMMAD SHOAIB
SEP HAZRAT UMAR
SEP ATTA UR REHMAN
SEP REHMAT SAHIB KHAN
C/SEP SHER AHMED
SEP SHYELI KHAN (3W)

greatness, favour, love of our creator, sitting at a small plateau overlooking the calm, green valley with River Chitral flowing with all the glaciated water. It was dawn with cool air blowing lightly and gently; Ramadan was coming to an end in another two days thus festivity of Eid was already in air and in the hearts of scouts. Many scouts had gone on Eid leave yesterday thus bringing the manning level at certain posts to unauthorised level.

Darashot is a small hamlet on the western bank of River Chitral, a small wooden hanging bridge allows the fifty odd inhabitants to cross over. On the other hand it allows the tribes, men, scouts to cross over and get into Afghanistan , area is thickly vegetated with olive trees, boulders, stones, defiles with odd patches of level field. Darashot was a small post and not a small piquet , it was a check point established at the eastern end of bridge, it was nothing more than a room with two tents pitched alongside having ammunition, living , cooking, for twenty scouts in them, commanded by Subedar Niat Ghazi. The main aim was to keep the wooden hanging bridges under observation, checking everyone physically who crosses it and at dusk to close it. A procedure being followed since 1901 in the area when ever there was any threat from Afghans. Historically River Chitral bifurcates the Hindu Raj like a peach line. Area on west was Kafiristan , extending from Arandu all along the west bank , reaching a conical omega bend at Nagur & Mirkhani, and encompassing Ayun, very few crossing points not more than a dosen are available.

Darashot was a complex of two posts, one which was at the bridge and other across the narrow track and higher on the plateau; in a small stone room which was built by the herdsman for self and animals protection. There were six scouts manning this out post, they had no line communication with the post down below. They daily had to go down to fetch the fresh ration or cooked food from the main post. Water was another issue, drinking water was also fetched from down below. Two sentries would give the duty at one time, but it was nothing alarming. There was one machine gun, one rocket launcher with six rounds, and four rifles at the post, it was a section strength weapons and ammunition.

' I heard a shot and I inquired from my comrade who was standing and looking at the vast, dimly lit frontier with weapon in his hand, I got up and walked towards him , that is when the first major burst of fire opened up, it included machine gun, with rocket launcher noise overpowering all other; they had hit the ammunition tent and now the crackling of rifle fire. It all took two minutes, now all of my comrades were out , running with the weapons and occupying the positions. None of us was sure as to what had actually happened and where should we fire in retaliation'[46].

Machine gun could not fire down below because of 'line of fire' which created a dead zone. In another few minutes situation became clear, own post down below was hit with rockets and small arms fire, enemy had established fire base on the western bank and now they were running across the bridge. Sepoy Nasir narrates

'I fired a rocket at the bridge to destroy it thus blocking the invasion but I missed it by few feet. Within seconds they were on our bank, now fire started coming on our locality too. There was no pause everything was happening in quick time'.

Apparently miscreants initially had very little or no information about the post at higher ground, it had affected their plan; they now ascended up and engaged the post. By 0710 hours, two scouts had embraced shahadat at the Darashot Post No.2 and almost all were dead at the main post down

46 Interview with Naik Nasr Minullah, October 2013.

below. By 0745 hours, the battle was going on at the higher post, miscreants apparently had achieved their aim but now wanted to wipe out the complex completely. One reason was that they had suffered casualties due to the fire coming from Nasir's post.

'In another hour I was the only one left alone at the post, around me were the dead bodies lying of my five comrades, I lost all sense of proportion, I became full of rage, I knew sooner or later a bullet is going to end my life like they taken away the lives of these five scouts. I was firing from different positions with different weapons which were lying around. I gradually started taking the dead bodies inside the room, I did not want them to be beheaded.'

Sepoy Nasir took the position inside the hut and relied upon the fire, soon two rockets came and hit his bunker, room was full of smoke and nothing was visible, two Taliban came rushing in following the rocket burst, Nasir saw just a silhouette and fired back, it hit the Taliban in head and he died instantly, now the battle took another dimension, Taliban were now eager and desperate to get the dead body back and Nasir was holding the post single handed. Duel continued for another hour, and then Taliban offered him safe passage if he allows them to take the dead body, Nasir fired a burst in reply. Taliban made another attempt to enter, destroy the hut but failed, by mid day Nasir was left with only seven bullets of rifle. At this point Taliban vacated the post and the area as rescue was approaching from Mirkhani. Nasir was ignorant about all this and when he heard the voices of men asking him to come out he refused and it was only when his own wing mate came and he recognised their voice, he come out.

Sepoy Nasir was promoted and then sent on United Nations Peace Corps in Sierra Leone. He has one more brother who is also serving in Chitral Scouts.

Orakzai Agency. A company strength ex Chitral Scouts was sent to Orakzai Agency in May 2010 for 'Operation Khawakh Bade Sham'. The company performed exceptionally well and repulsed a number of attacks of the miscreants inflicting heavy casualties to them while defending 'Top' in Daburi. On night 21/22 June 2010, the company repulsed the major miscreants attack causing them 43 dead and 56 injured. Own losses were four shaheed (including two from Punjab Regiment) and sixteen injured.

					Rustum Mardan, mcts atk QRF, he received bullet injury and embraced shahadat.		
9.	404624	Lnk	Qadir Khan		The indlwas performing duty at Tutano Banda post Swat, embraced shahadat due to mcts sniper fire on 27 Oct 2008.	27-10-2008	01
10.	405978	"	Bashir Ahmad	"	While performing duties ub Ops Al-Mizan Rah-e-Rast at Dir Maidan, when cvy left for Lal Qilla Dir Maidan, mcts atk cvy at Haya Sari and opened fire. Resultantly the indl embraced Shahadat.	26-4-2009	01 " ✓
11.	406359	"	Shah Hussain	"	While performing duties ub Ops Al-Mizan Rah-e-Rast at Dir Maidan Tank was broken down. Coy was moveing towards the incident for escape the manpower. Suddenly mcts atk on QRF, resultantly the indl embraced Shahadat.	26-4-2009	01 " ✓
12.	405419	Sep	Ajaib Khan	"	While performing duties ub Ops Al-Mizan Rah-e-Rast at Dir Maidan, when cvy left for Lal Qilla Dir Maidan, mcts atk cvy at Haya Sari and opened fire. Resultantly the indl embraced Shahadat.	26-4-2009	01 " ✓
13.	406115	"	Bashir Ahmad	"	While performing duties ub Ops Al-Mizan Rah-e-Rast at Dir Maidan, when cvy left for Lal Qilla Dir Maidan, mcts atk cvy at Haya Sari and opend fire. Resultantly the indl embraced Shahadat.	25-4-2009	01 " ✓
14	406721	"	Aman ur Rehman	"	While performing dutíes ub Ops Al-Mizan Rah-e-Rast at Dir Maidan, when cvy left for Lal Qilla Dir Maidan, mcts atk cvy at Haya Sari and opend fire. Resultantly the indl embraced Shahadat.	26-4-2009	01 " ✓
15	406834	"	Sher Nawaz	"	While performing duties ub Ops Al-Mizan Rah-e-Rast at Dir Maidan, when cvy left for Lal Qilla Dir Maidan, mcts atk cvy at Haya Sari and opend fire. Resultantly the indl embraced Shahadat.	26-4-2009	01 " ✓
16	406496	"	Sher Azam	"	While performing duties ub Ops Al-Mizan Rah-e-Rast at Dir Maidan, when cvy left for Lal Qilla Dir Maidan, mcts atk cvy at Haya Sari and opend fire. Resultantly the indl embraced Shahadat.	26-4-2009	01 " ✓

17.	406276	Sep	Bahadar Shah	Ctl	While performing duties ub Ops Al-Mizan Rah-e-Rast at Dir Maidan, when cvy left for Lal Qilla Dir Maidan, mcts atk cvy at Haya Sari and opend fire. Resultantly the indl embraced Shahadat.	26-4-2009	01 " ✓
18.	405557	"	Sajjad Hussain	"	Embraced Shahadat at (Kalaya) Orakzai Agency due to IED blast.	28-05-2010	"
19.	400229	"	Nadir ud Din	"	Embraced Shahadat at FCNA due to latomyosarcoma.	03-05-1993	
20.	406361	"	Shahzad Rahim	"	The indl was performing duty at Devloi Post Swat, embraced Shahadat due to mcts Rkt fire on 19 Jun 09 at about 2025 hrs.	19-6-2009	01 "
21.	403588	"	Gul Zahid	"	The indl was performing at Kmarsar-II Post Bajaur, embraced Shahadat due to mcts attack at Kmarsar-II Post Bajaur on Nl 01 Dec 09 at about 0145 hrs.	01-12-2009	01 "
22.	404151	"	Mahboob Khan	"	Embraced Shahadat due to en Arty shelling in op at FCNA Koh-i-Paima at badar Base on 3 Jul 1999.	03-7-1999	01 "
23.	405718	"	Wali Khan	"	The indl was performing duty at Ayub Bridge Post Swat, embraced Shahadat due to veh suicidal attack on 6 Nov 2008.	06-11-2008	01 "
24.	407750	"	Noor Ahmad	"	The indl was performing duty at Mandal Nullah Post Bajaur, embraced Shahadat due to IED blast during patrolling at Mandal Nullah Post on 24 Feb 2010 at about 0800 hrs.	24-2-2010	01 "
25.	405167	"	Ali Hazrat	"	Embraced shahadat due to en Arty shelling in op at FCNA Koh-i-Paima at badar Base on 30 Jun 1999.	30-6-1999	01 "
26.	[illegible]005	"	Noor Khan	"	FCNA Embraced Shahadat due to en Air attack at Koh-i-Paima at badar Base on 01 Jul 1999.	01-7-1999	01 "

Chitral Scouts under the directive of Frontier Corps, constructs accommodation and provide it free of cost to the next of kin of Shaheeds. Jobs and other help is also extended to them.

آغاز منصوبہ بحکم جناب
میجر جنرل غیور محمود تمغہ بسالت
انسپکٹر جنرل فرنٹیئر کور صوبہ خیبر پختونخواہ
تکمیل منصوبہ بتاریخ: 20 اکتوبر 2012ء
آغاز منصوبہ بتاریخ: 7 جولائی 2012ء
تعمیر کنندہ یونٹ: 2 ونگ چترال سکاؤٹس
CHITRAL SCOUTS

Chapter Seven

History of Chitral Scouts Wings

The basic organization of a fighting unit in Chitral Scouts is a wing which is equivalent to a light infantry battalion commanded by a lieutenant colonel. It has a bayonet strength of 615 men. Wing is divided into four companies, each company having three platoons, each platoon having three sections, strength of section is ten men. Wing has its own firepower in the form of mortars, it also have signal, military transport and medical platoons.

1 Wing:

Senior most wing of Chitral Scouts, it was known as No.13 Wing and in 1990 was reverted to its old and present designation. Veteran of 1919 and Kashmir Wars. In June 1999 it entered the operational area in Kargil. Since 2000, The wing was deployed in Dir Maidan operation where it was placed under command Dir Scouts since September 2009.

2 Wing

It was an integral part of Chitral Scouts, in 1989 its designation was changed from No.14 Wing to No.2 Wing. It has served at Arandu, Warsak, Mirkani, Drosh and Chakdara. Wing has also served at FCNA, {Chilas 1993, Piun, September 1994,Hamzagond May 1999 and Piun June 2002}. In June 2007 No.2 Wing replaced No.3 Wing at Swat, headquarters were established at Kabal. From here the scouts moves to Tutain Banda with a Baluch Regiment, the post was almost attacked daily by the snipers which took the life of Lance Naik Qadir Khan and seriously wounded Sepoy Tahir Shah. Apart from snipers the post was subject to suicide attack as well. Post commander Subedar Muhammad Daud and Naib Subedar Shehr Gulab displayed extra ordinary display of leadership and kept the morale high. Another volatile post was at Ayub Bridge , it was attacked by an explosive laden suicide vehicle which caused shahadat of Sepoys Wali Khan, Haji Rehmat and Imtiaz with further ten more scouts getting wounded. Later the wing was move to Bajaur. In October 2010, the wing moves back to Mirkhani. (Chitral).

3 Wing:

No.3 Wing was raised on 1st October 1986 and re-designated as No.3 Wing in 1989. Wing has served under FCNA in August 1991-1992, May 2000- June 2001; from where it was rushed to Kitkot in Bajaur. During the Pakistan-India escalation in 2001, the Wing remained deployed at Baddomallhi in Punjab for over three months. Earthquake relief operation in 2005 was actively participated by the wing at Shangla, Ghari Habib Ullah and Manshehra. 3 Wing was ordered to move to Swat on a very short notice in September 2007; initially wing was deployed in Mingora and surroundings. Later wing was deployed within area of responsibility of army infantry brigade while wing headquarters was established at Kabal. Security of two main bridges (Ayub and Shamozai bridge} was the responsibility of wing. Besides number of cordon and search operations, one suicidal vehicle was also identified and destroyed with bombers on 18 December 2007. The wing is in Bajaur Agency operation under command

Bajaur Scouts since Oct 2010.

4 Wing:

The wing was raised in 1986 as No.16 Wing and later re-designated to present in 1989. In 1992 the wing was placed under command FCNA for two years where it served at Siari, Olding and Hamzigund sectors. A year of internal security at Gilgit in 1998 and another at Bajaur in 1999 before taking over the Mirkani Fort from No.5 Wing in April 2001. In May 2003 the Wing again move into FCNA for a year. Wing has been in Milward (Khyber Agency) and performed operational duties in a very befitting manner from September 2007 – October 2009. Recently the Wing is performing duties in Mastuj Sector (Chitral). A company (-) ex this wing is still deployed in Bajaur.

5 Wing:

The wing performed their duties in Swat and Dir Maidan Operation very well. This wing started advance on 26 April 2009 from Timergara and cleared area up to Lal Qila, heavy exchange of fire took place with miscreants and eight scouts including one junior commissioned officer embraced shahadat on same day. Presently the wing is performing their duties in Drosh.

6 Wing

6 Wing was raised on 1st September 1989 at Drosh with Major Afzal as the first wing commander. They were stationed at Chakdara before taking part in Bara operation {1990} followed by Malakand Opeartion {1991}. It served in FCNA from August 1997 – September 1998 and had another tour of duty in the same operational area in June 2006. After two years of stay at Mastuj it was inducted in Dara Adam Khel in 2009 and remained actively deployed till May 2012.

CHITRAL SCOUTS
WING COMMANDERS
1 WING
FROM TO
MAJOR KHALID PERVIEZ FF AUG 91 FEB 93
CAPT MUKHTAR AHMED FEB 93 APR 93
CAPT WAQAR AZIZ ALVI FF APR 93 JUN 93
MAJOR SOHAIL AHMED JUN 93 SEP 94
MAJOR ABDULLAH MEHMOOD BALOCH SEP 94 JAN 97
CAPT YAHYA MAY 97
LT COL SEP 97
MAJOR AHSAN SALEEM BALOCH OCT 97
MAJOR NASIR AIR DEF FEB 98
LT COL KALEEMUD DIN ARTY NOV 98
LT COL ATHAR UR REHMAN ARTY OCT 99
LT COL MUHAMMAD KAUSAR ARTY DEC 99 AUG 2000
MAJ ABDUS SABOOR ARTY AUG 2000 OCT 2000
MAJ ABDUL NASIR AHSAN ARTY OCT 2000
MAJ AIR DEF
MAJ DAUD AHMAD
MAJ SYED ALI IMRAN SHAH AC
MAJ MUHAMMAD IQBAL FF
MAJ ARTY JUN 2005
MAJ FARHAN MUFTI BALOCH JUN 2005 APR 2006
MAJ SAJID ABRAR SIGS APR 2006 OCT 2006
MAJ SIND SEP 2007
MAJ NOV 2008
LT COL BALOCH SEP 2010
LT COL ARTY
WING COMMANDERs NO.1 WING CS
LT COL MUHAMMAD IMAD AZIM ARTY SEP 2010 TO MAR 2013
LT COL NAVEED NAZAR AWAN ARTY MAR 2013 TO
SHOAIB
FAYAZ

SUBEDAR MAJORS
1 WING
FROM TO
SM AHMED JAN TK I 01 3 33 23 2 86
SM GUL MUHAMMAD 31 7 86 31 8 88
SM KHURSHID NIZAR BAIG 01 9 88 30 9 89
SM ABDUL GHAFFAR KHAN 01 10 89 30 9 90
SM AHMED WALI 01 10 90 14 1 91
SM MUHAMMAD ZAFFAR TK II 15 1 91 19 7 92
SM ALI NAWAZ 20 7 92 03 10 93
SM QAMAR UD DIN 04 10 93 14 12 94
SM MUSTAFA KAMAL 15 12 94 31 10 96
SM MUHAMMAD GHAFFAR KHAN 01 11 96 01 3 98
SM GHULAM RASUL 14 8 98 22 1 2000
SM SARDAR WALI 22 1 2000 24 5 2002
SM HAKIM SARWAR 24 5 2002 30 11 2002
SM QURBAN ALI 01 12 2002 28 6 2003
SM MAHBOOB ALAM 28 6 2003 10 8 2004
SM NASIR AHMED 10 8 2004 25 6 2006
H/CAPT SYED MUSA WALI SHAH TK I 25 6 2006 14 8 2007
SM ABDUL NASIR 15 8 2007 01 8 2008
SM ABDUL ALIM 01 8 2008
SM SAEED HASSAN TK II 15 8 2011

LNK ALI MURAD
DEATH

CHITRAL SCOUTS
SUBEDAR MAJORS 2 WING
NAME
FROM
TO
ABDUL HAMID
16 AUG 2000
HAKIM SARWAR
30 APR 2002
SARDAR WALI
2 WING UNDER FCNA323BDE
FROM10MAY 02 TO 2AUG 03
ABDUR RAHIM BAIG
23.3.2005
ABDUR RAHIM BAIG
14 9 2005
SYED SALAH UD DIN
14-9-2005
ZARFARAZ SHAH
04 8 2007
SYED ARAB SHAH
14 8 2010
CHITRAL SCOUTS
2WING COMMANDERS
MAJ ABDUL SHAKOOR AK APR
LTCOL QAZI NAJEEB HAIDER AD AUG
MAJ M.ALI HYDERI(SHAHEED) BR SEP
LTCOL M. ABDUL GHAFOOR BR DEC
CAPT KHALID SAEED BR MAY
MAJ SODAGAR HUSSAIN SR JUL
MAJ DAUD AHMAD SR APR
MAJ ABDUL NASIR IHSAN ARTY OCT
LTCOL SYED NADIR ALI SHAH BR DEC
CAPT SYED KAMRAN ALI SHAH PR AUG
MAJ MINHAJ UD DIN AD AUG
MAJ MALIK IRFAN IJAZ AC OCT
LTCOL TAHIR NADEEM AC SEP
LTCOL M. KHALID RANA ARTY AUG
LTCOL NASIR LATIF FF SEP
LTCOL TOUQEER AHMED MALIK PR MAR

SUBEDAR MAJORS
1.SM MUHAMMAD ISMAIL TKI WEF 01_01_87 TO 31_10_88
2.SM HAJIBUR REHMAN WEF 01_11_88 TO 31_07_90
3.SM KHURSHID NIZAR BAIG TKI WEF 01_08_90 TO 30_11_90
4.SM MUHAMMAD ZAFFAR TKI WEF 01_12_90 TO 31_01_91
5.SM MUHAMMAD WALI WEF 01_02_91 TO 31_08_91
6.SM ABDUL GHANI WEF 01_09_91 TO 31_12_93
7.SM HAKIMUD DIN WEF 01_01_94 TO 30_07_95
8.SM MUHAMMAD AKBAR SHAH TKI WEF 01_08_95 TO 30_09_95
9.SM MUHAMMAD AYUB TKI WEF 01_10_95 TO 14_08_98
10.SM ABDUL HAMID WEF 15_08_98 TO 12_10.2000
11.SM GULAB NIZAR WEF 12_10_2001 TO 21_04.2001
12.SM SHER AFZAL WEF 14_08.2001 TO 15.11_2002
3.SM AZMAT ULLAH WEF 15_11_2003 TO 01.12_2003
4.SM AMIR WALI KHAN WEF 18.12_ 2003 TO 25-7-2004
15.SM MAHBOOB ALAM TKI WEF 10.8 2004 TO 14. 5.2006
16.SM MUHAMMAD ULLAH JAN WEF 15_5_ 2006 TO 01 08 2008
17.SM ABDUL SAMAD TK1 WEF 01_08.2008 TO 04.12.2009
18.SM MUHAMMAD AJAB KHAN WEF 01_03.2010 TO 28.02.2013
19 SM FATEH ULLAH WEF 01.03.2013
3 WING
WING COMMANDERS
1 MAJOR MUHAMMAD MOAZZAM (ARTY) WEF SEP 94 TO DEC 95
2 MAJOR KHALID MAHMOOD (FF) WEF DEC 95 TO NOV 96
3 MAJOR QUTAB.U.DDIN (ARTY) WEF NOV 96 TO JAN 97
4 MAJOR AHSAN SALEEM (BALOCH) WEF JAN 97 TO SEP 97
5 MAJOR MUHAMMAD NAVEED KHAN (BALOCH) SEP 97 TO NOV 97
6 MAJOR NIZAM.UD.DIN (PUNJAB) WEF NOV 97 TO JUL 99
7 MAJOR SODAGAR HUSSAIN (SINDH) WEF JUL 99 TO JUN 2000
8 LT COL MUHAMMAD ABDUL GHAFOOR Isc (BALOCH) WEF JUN 2000 TO 2001
9 MAJOR MUHAMMAD KHALID (SINDH) WEF JUN 2001 TO JUL 2001
10 MAJOR DAUD AHMAD (SINDH) WEF JUL 2001 TO MAR 2002
11 MAJOR MUHAMMAD KHALID (SINDH) WEF MAR 2002 TO JUL 2002
12 CAPT MUHAMMAD IMRAN BABAR AK WEF JULY 2002 TO JAN 2003
13 CAPT MUHAMMAD SHARIF-UD.DIN (BALUCH) WEF JAN 2003 TO AUG 200
14 CAPT FAROOQ SHAH (BALOCH) WEF AUG 2003 TO SEP 20
15 MAJOR ZULFIQAR AHMED gsc (ARTY) WEE SEP 2004 TO SEP 200
16 MAJOR MUHAMMAD ASHRAF MALIK (SIG) WEF SEP 2006 TO SEP 200
17 LT COL RAJA ZAFAR KHAN (FF) WEF SEP 2007 TO AUG 20
18 LT COL SAJID MAHMOOD (ARTY) WEF AUG 2009 TO OCT 2011
19 LT COL SAGHEER ARSHAD (ARTY) WEF OCT 2011 TO SEP 2012
20 LT COL NAEEM ASIF (ARTY) WEF SEP 2012
CHITRAL SCOUTS

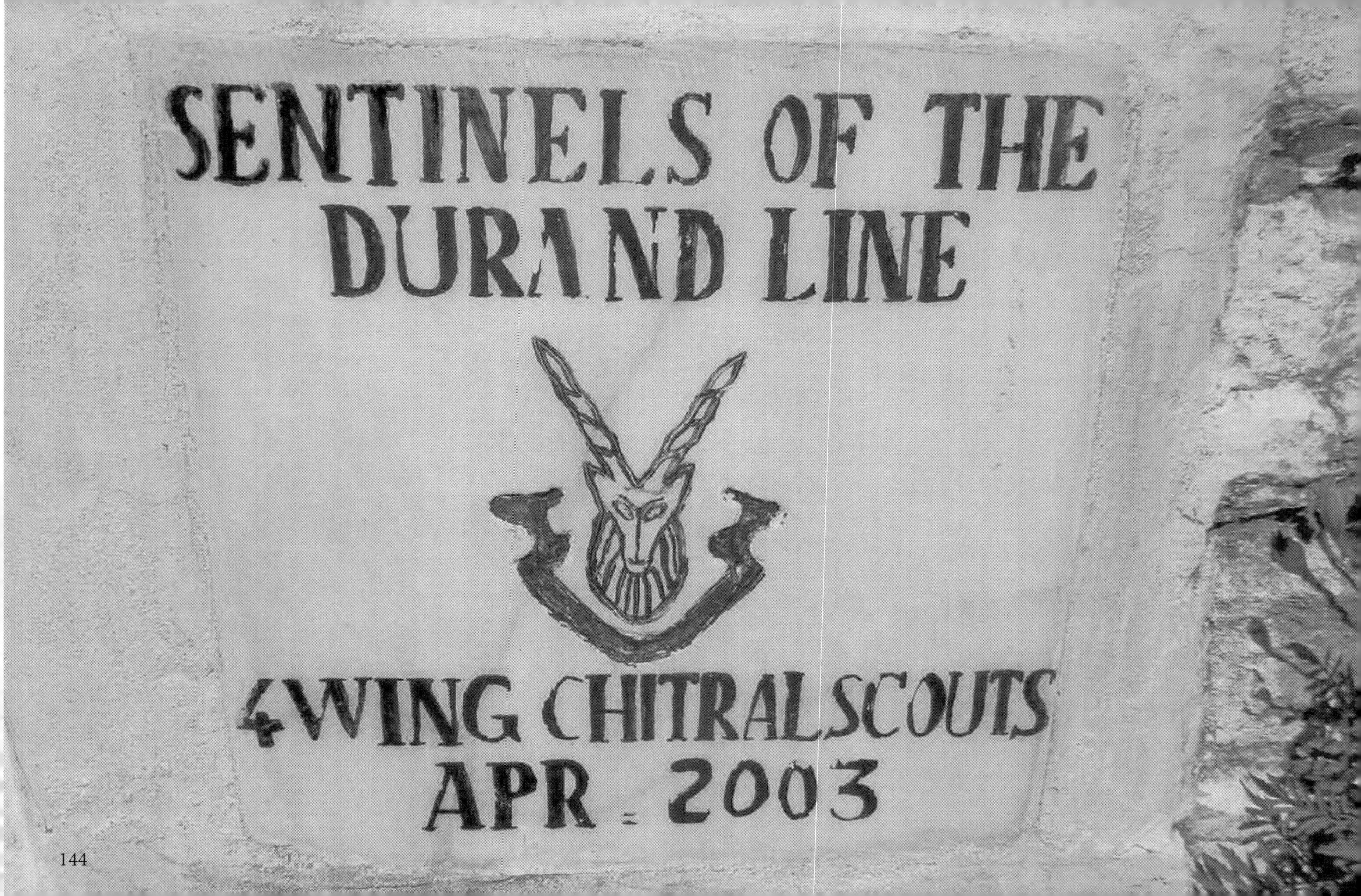
SENTINELS OF THE
DURAND LINE
4WING CHITRAL SCOUTS
APR - 2003

CHITRAL SCOUTS

یا اللہ
۵
چترال
یا قیوم
یا غفار
یا رحیم

Wing Commander Lt Col Afraz

Subedar Major Miftah

No. 6 Wing

IHSAN
KHALIL

مارٹر بیٹری
مارٹر بیٹری کے فائیر پر ایک نظر
MOR BTY
BTY COMD
PSS 11685 MAJOR KHURSHID MUSLIM FEB 90 TO JUL 90
PSS 14682 MAJOR M JAVID IQBAL JUL 90 TO SEP 90
PSS 15063 MAJOR RANA AFTAB LATIF SEP 90 TO DEC 90
PA 21132 MAJOR M. ADNAN MAQSOOD OCT 92 TO SEP 93
PA 23304 MAJOR MUHAMMAD MUAZZAM NOV 93 TO SEP 96
PSS 23535 MAJOR MUNAWAR HUSSAIN SEP 96 TO SEP 96
PA 28455 CAPT KAMRAN ALI OCT 96 TO DEC 96
PA 26075 MAJOR ARSHAD AHMAD JAN 97 TO APR 97
PA 23091 MAJOR NASIR ALI MAY 97 TO JULY 97
PA 23979 MAJOR IRSHAD HUSSAIN NASIR AUG 97 TO SEP 97
PA 28455 CAPT KAMRAN ALI OCT 97 TO JUNE 98
PA 32163 CAPT SALMAN ANWAR JULY 98 TO MAY 99
PA 26661 MAJOR SHER AFZAL MAY 99 TO DEC 99
PSS 23535 MAJOR SYED AKMAL HUSSAIN DEC 99 TO DEC 2002
PA 30073 MAJOR SAJID SARWAR JAN 2003 TO MAY 2005
PA 30744 MAJOR SHAHID KARIM KHAN JUNE 2005 SEP 2008
PSS 26392 MAJOR MUHAMMAD AKBAR SEP 2008 TO DEC 2009
PA 32826 MAJOR SALMAN JAMAL BUTT JAN 2010 TO SEP 2010
PA 32526 LTCOL SALMAN JAMAL BUTT OCT 2010 TO OCT 2012
3.7 HOW MK
1943
BRITISH ORIGIN
RANGE 6200M
RDS FIRED- 652
LAST FIRED IN CS
10 JUN 1987

WELCOME TO THE HOME OF
CHITRALI GUNNERS

Chitral Scouts Hospital - Chitral

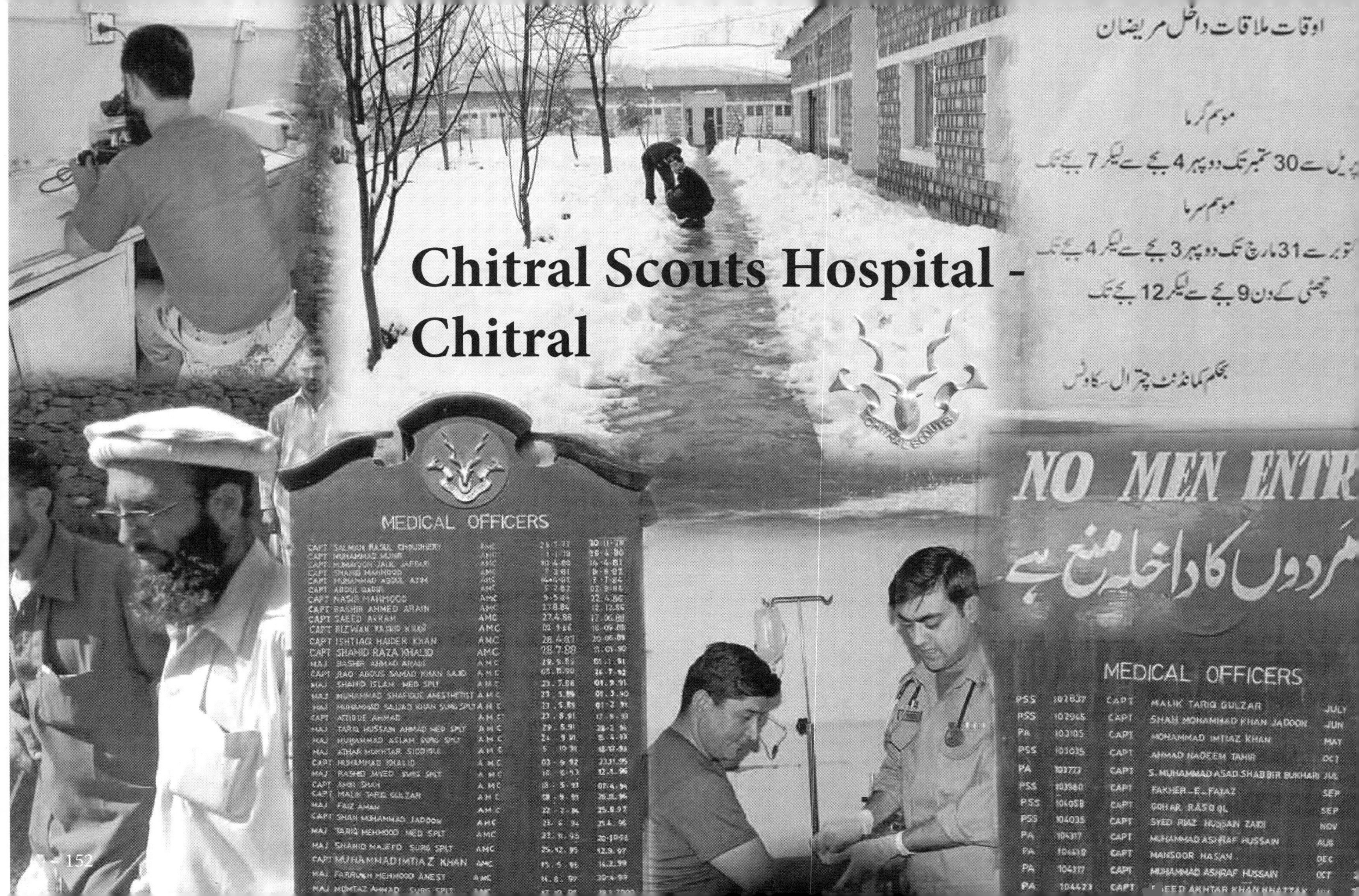

SCOUTS HOSPITAL CHITRAL
INAUGURATED
BY
GEN PERVEZ MUSHARRAF NI(M)TBt
CHIEF OF ARMY STAFF
ON03 SEP1999
CHITRAL SCOUTS HOSPITAL CHITRA
چترال سکاؤٹس ہسپتال چترال
SURGICAL SPECIALISTS
PSS 102660 MAJ AFTAB AHMAD 1.1.2000 1.10.2002
P&S 102861 MAJ SHAHID IQBAL 6.9.2002 25.1.2003
PA 103269 MAJ ABDUL RASHID 24.12.2003 12.10.2007
PSS 103712 MAJ KHALID MEHMOOD 29.09.2007 14.12.2009
PA 103355 MAJ M.ASIM NIAZ 15.12.2009 25.07.2012
PA 104385 MAJ SHAFAQAT ALI 26.07.2012

FRONTIER CORPS PUBLIC SCHOOL DROSH
F C PUBLIC HIGH SCHOOL DROSH
PRINCIPALS
CAPT SAJJAD HUSSAIN MA ISL JUL97
MAJ SODAGAR HUSSAIN MA PS SEP00
MR ALI NAWAB MA ENG, BEd, BLSc SEP 2000
MR GHULAM WAZIR
MR JUL 07
(IN THE MEMORY OF SHUHADAS OF 5 WING CS
RENOVATION AND CONSTRUCTION OF JUNIOR
WING FCPS DROSH
CARRIED OUT BY
WING COMD LT COL SALMAN JAMAL BUTT
N/SUB. SHAHABUD DIN
HAV, SHER
AND ALL RANKS 5 WING CS REAR SC
FROM 2011 TO SEP 2012
THE ONLY GOOD IS KNOWLEDGE
AND THE ONLY EVIL IS IGNORANCE
CAR PARK
A TEACHER IS A
CHIEF ACTOR IN
THE DRAMA OF LEARNI
DATE
TOTAL
PRESENT
ABSENT
LEAVE
HOSTEL
FRONTIER CORPS PUBLIC SCHOOL
F C PUBLIC HIGH SCHOOL DROSH
ROLL OF HONOUR
S NO SESSION NAME R/NO OB/MARKS GRADE
1. 2003 ABDUL SABOOR 534/850
2. 2004 584/850
3. 2005 MOHSINULLAH 115265 A ONE
4. 2006 SHAH FAHAD UL MULK 117015 970/1050 A ONE
5. 2007 YASIR ALI HAMZA 717/850 A ONE
6. 2008 TABRAIZ WALI SHAH 716571 691/850 A ONE
7. 2009 IJAZ UL HAQ
8. 2010 FAZLE MAULA 116709
9. 2011 SOHAIL NAWAZ A. ONE
10. ATTAULLAH

Chitral Scouts Public Schools

English Medium Public Schools under the name of Army Public School were established at Drosh and Chitral during February 1992 and March 1997 respectively which have now been re-designated as Frontier Corps Public School from 1st June 2003. Presently the schools are providing education from play group to 10th class. However ample space has been provided to cater for increasing needs of future on its subsequent up-gradation. These schools are self-finance public schools and following the curriculum of Army Public School and functioning under the umbrella of APSACS (Army Public School College System) secretariat. The school are being on no profit no loss system. Boarding facility is being provided at Chitral and Drosh to accommodate the children of shaheed and wounded scouts. There are 800 children studying at Chitral, school have 20 civilian and eighteen teachers from Chitral Scouts imparting education. At Drosh the strength is 650 students with 24 teachers, half are civilians and rest are from Chitral Scouts education wing.

Sadqa, it is performed every month

Signal Platoon

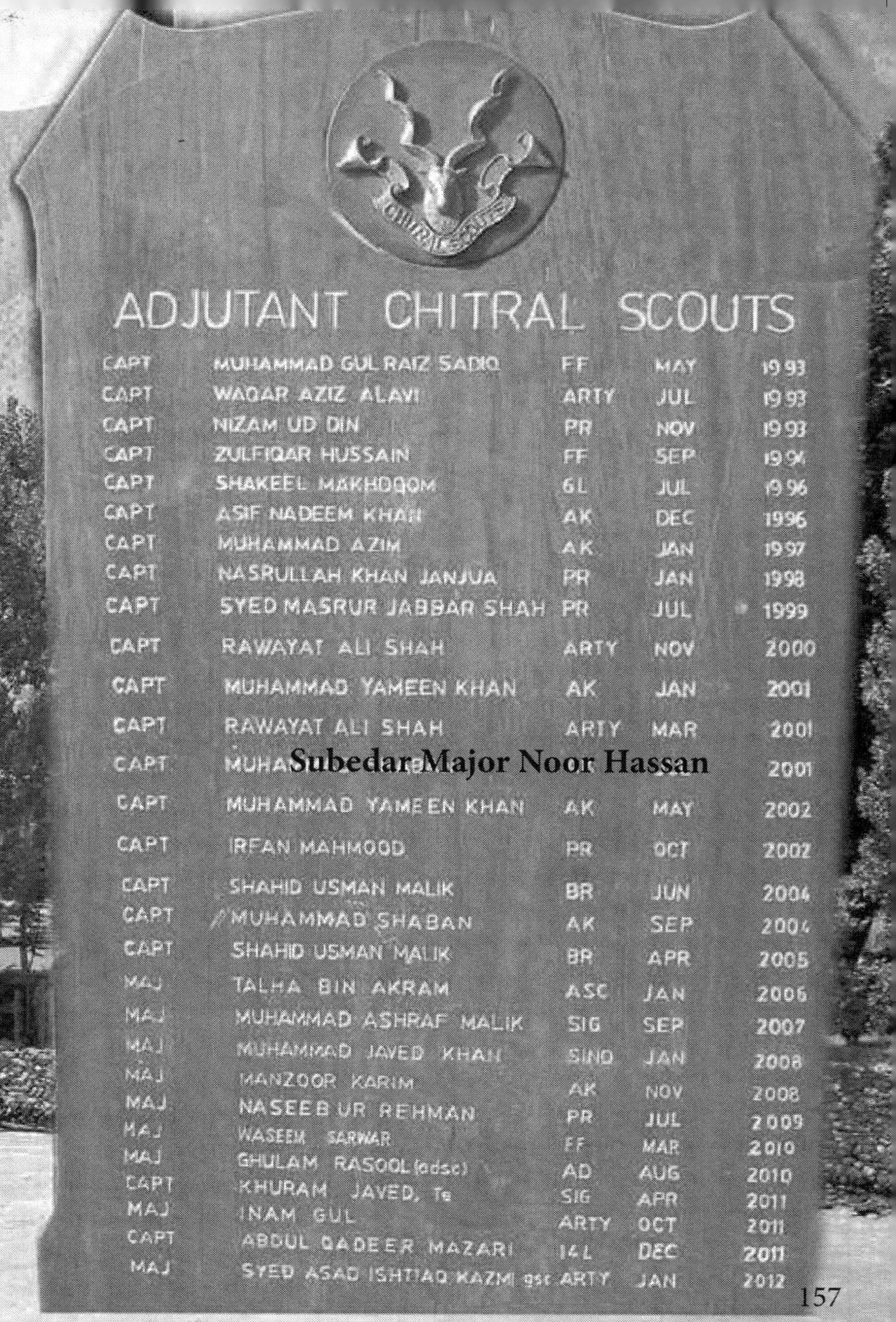

Subedar Major Noor Hassan

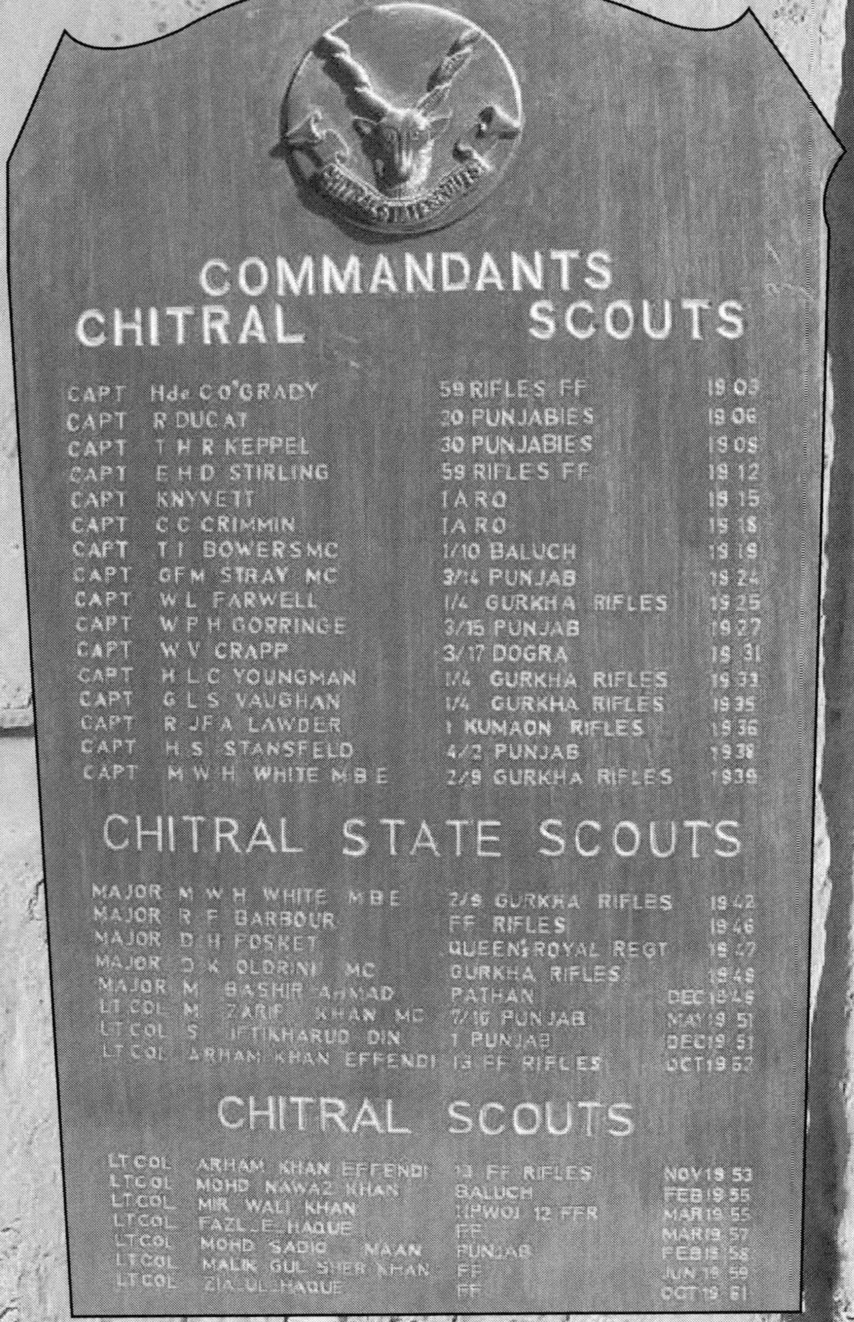

Grady

Lt Col M Zarif Khan, MC

Lt Col S. Iftikhar Ud Din

Lt Col Mir Wali Khan

Lt Col Malik Gul Sher Khan

Lt Col Zia Ul Haq

Lt Col Shoukat Sultan

Lt Col M.Sher Khan, SJ

Lt Col Fahimullah Khan Khattak

Lt Col Muhammad Afzal Khan

Lt Col Basharat A.Sultan,SJ & Bar

Lt Col Murad Khan Nayyar

Col Hameed Ullah Pawar

Col Akhtar Zamin

Col Syed Amjad Mukhtar Shah

Col Javed Kamal Meyer

Col Faheem Anwar

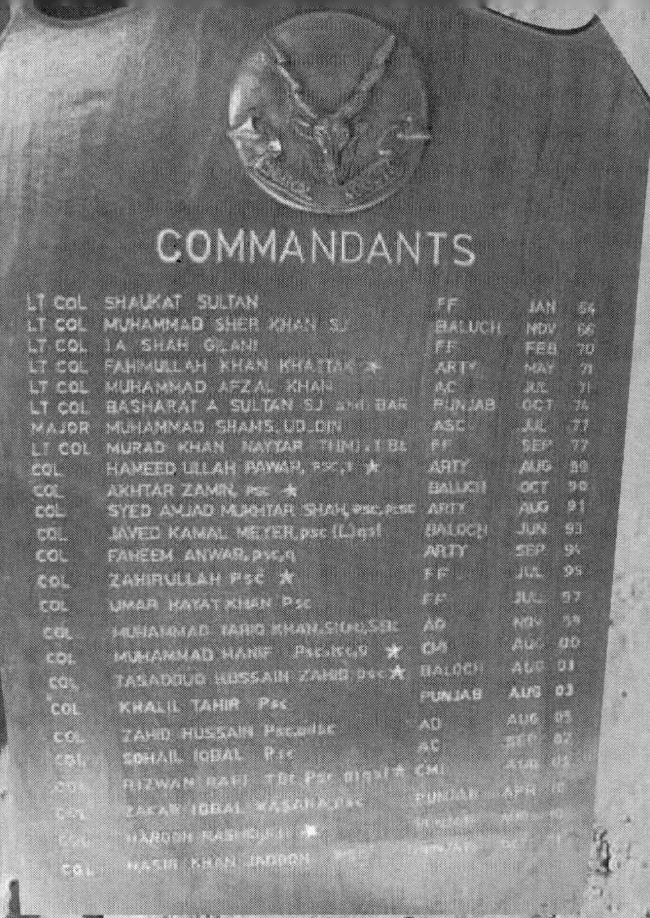

Col Zahirullah

Col Umer Hayat

Col Muhammad Tariq Khan

Col Tasadduq Hussain Zahid

Col Khalil Tahir

Col Zahid Hussain Awan

Col Sohail Iqbal
16 Sep 2007 - 16 Aug 2009

Col Rizwan Rafi

Kassana

Haroon

Jadoon

SUBEDAR MAJOR
CHITRAL STATE SCOUTS
SUB MAJ ABDUL RAUF JUN 42
SUB MAJ HABIB UL AHMAD JUN 48
CHITRAL SCOUTS
SUB MAJ JAN BADSHAH TKII JULY 53
SUB MAJ BADUR UL MULK TK II OCT 60
HONY CAPT MOMIN UL MULK TK I NOV 63
HONY CAPT MUHAMMAD SULAIMAN KHAN TK I SEP 67
SUB MAJ SHUKR-L AMIN TK I SEP 70
SUB MAJ SYED QASIM JAN SEP 73
SUB MAJ ZULFIQAR UL MULK SEP 76
SUB MAJ MUNSI KHAN TK I APR 80
SUB MAJ RAHIM FIDA TK II APR 83
SUB MAJ NOOR AHMAD AUG 85
SUB MAJ GUL MUHAMMAD TK I APR 88
SUB MAJ ABDUL GHAFFAR KHAN TK I AUG 89
SUB MAJ KHURSHID NIZAR BEG TK II SEP 90
SUB MAJ AHMAD WALI KHAN TK II DEC 91
SUB MAJ MUHAMMAD ZAFFAR KHAN TK I AUG 92
SUB MAJ ALI NAWAZ SEP 93
SUB MAJ MUSTAFA KAMAL JAN 94
SUB MAJ HAJI MUHAMMAD TK II MAR 94
SUB MAJ HUSSAIN JAN TK II OCT 94
SUB MAJ QAMAR UD-DIN TK II AUG 95
SUB MAJ MUSTAFA KAMAL TK I SEP 96
SUB MAJ NOWROZ KHAN TK I JAN 97
SUB MAJ M. ZAFFAR ALI SHAH MAY 97
SUB MAJ MUHAMMAD AKBAR SHAH TK I DEC 97
SUB MAJ MUHAMMAD GHAFFAR KHAN TK I AUG 98
SUB MAJ SHAH WALI SEP 99
HON CAPT GHULAM RASOOL TK1 JAN 2000
SUB MAJ AFAS KHAN TK1 APR 2001
SUB MAJ GULAB NIZAR SEP 2002

SUBEDAR MAJOR
CHITRAL SCOUTS
HON LT REHMAT HUSSAIN TKII MAY 2001
HON CAPT SAHIB NIZAR TK I DEC 2003
HON LT AZMAT ULLAH TKII AUG 2004
HON CAPT AMIR WALI KHAN TK I NOV 2006
HON CAPT GHULAM RASOOL TK I AUG 2008
HON CAPT ABDUL NASIR TKI DEC 2009
HON CAPT MUKHTAR ALI KHAN TKI JUL 2010
HON CAPT MIR BAHADER TKI AUG 2011
HON LT KAKA JAN TKII DEC 2012
HON LT NOOR WALI TKII JUL 2013
SAEED HASSAN

DATE	NAME	ADDRESS
June 85	General M. Zia-ul-Haq	It has been a great pleasure to be with Chitral Scouts again. It is a fine unit which Lt Col Murad commands. I wish them all the best. Zia-ul-Haq

DATE	NAME	ADDRESS
July 7, 1994	Benazir Bhutto I would like to thank the Chitral Scouts for the excellent arrangements made for the Polo Shandur match. The foreign dignitaries and Pakistani guests were looked after down to an excellent lunch of chapli kebabs and desi bread.	The Prime Minister House, Islamabad

DATE	NAME
July 93	General Abdul

DATE	NAME
9 Jan 81	Mughal Gen.

DATE	NAME	ADDRESS
Sep 99	Gen Pervez Musharraf COAS.	Enjoyed my stay at the Chitral Scouts Mess thoroughly. Their warm hospitality will remain memorable.

ADDRESS

DATE

NAME

COAS

It has been exciting to land at Dahireh unannounced and be exposed to the breathtaking beauty of your valley and, more so, to your warm hospitality

Thanks indeed

Apr 96

Khuhnawgt-ul-Mulk
Very many thanks for the wonderful hospitality to my friends a

Masty Chitral Dist

Lt.-Colonel Ian Keelan (Retd) Royal Engineers — 43 Audley Road, SAFFRON WALDEN

My father-in-law, H.C. EEN SANDEMAN OBE, RE, as Lieut Bengal S & M, designed and built the (motor) suspension bridge by the (1895) Fort, in 1925. This mostly journey was to inspect it and report back to him, now aged 95.

We have been royally welcomed and very generously treated by Chitral Scouts and the Comd. Col Zaharulla.

Thank you!

Ian Keelan

Hugh Keelan. Vermont, USA.

Many thanks for a wonderful

Hazel Keelan,

Thank you - and I look forward to seeing your new mess

119 GOLDHAWK R

LONDON W12

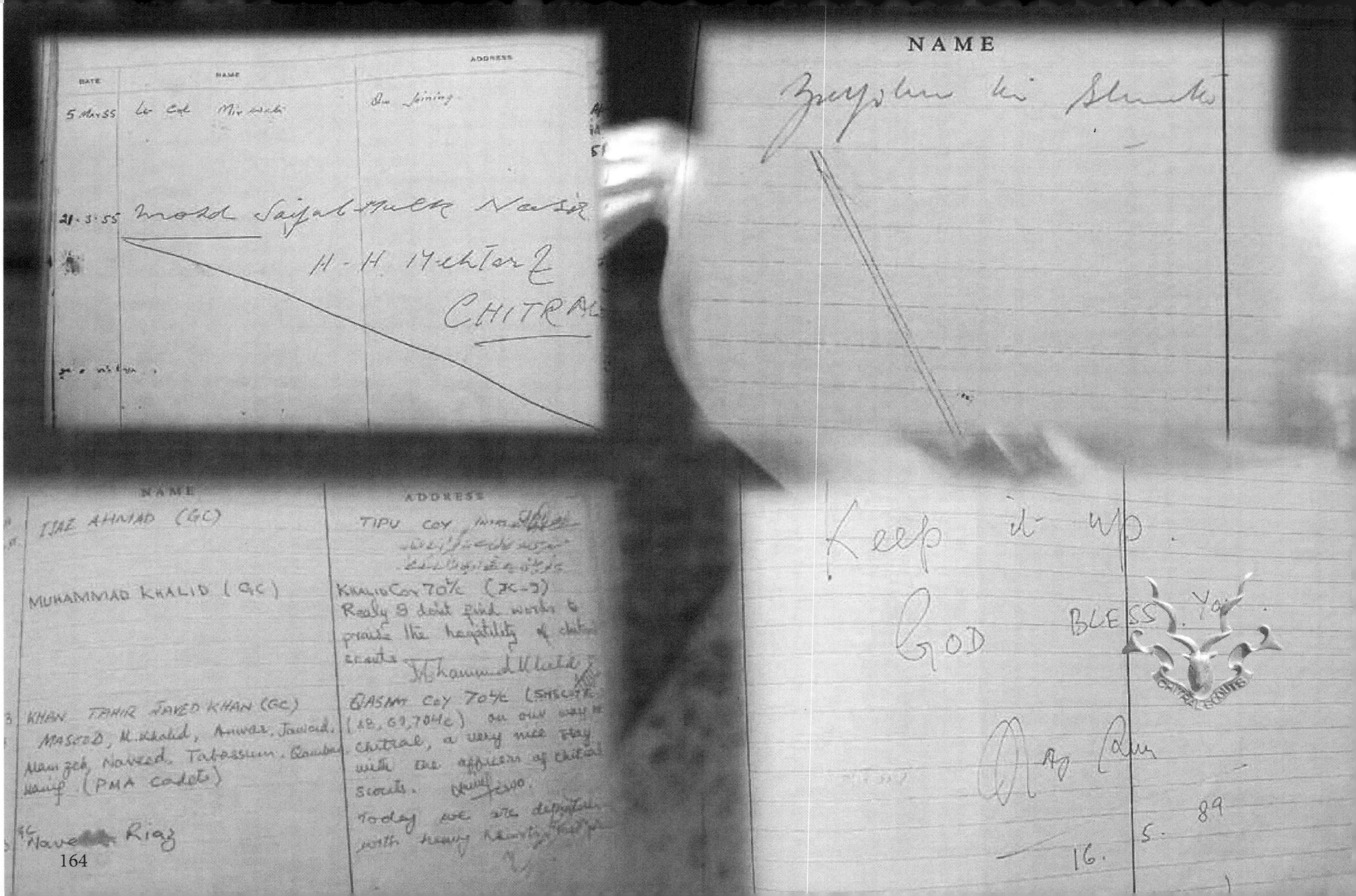
ADDRESS
DATE
NAME
5 Mar 55
21.3.55
H.H. Mehtar of
CHITRAL
NAME
NAME
ADDRESS
MUHAMMAD KHALID (GC)
Really I don't find words to praise the hospitality of Chitral scouts.
KHAN TAHIR JAVED KHAN (GC)
(PMA cadets)
on our way to chitral, a very nice stay with the officers of Chitral Scouts.
Keep it up.
GOD BLESS You.
16.5.89
CHITRAL SCOUTS

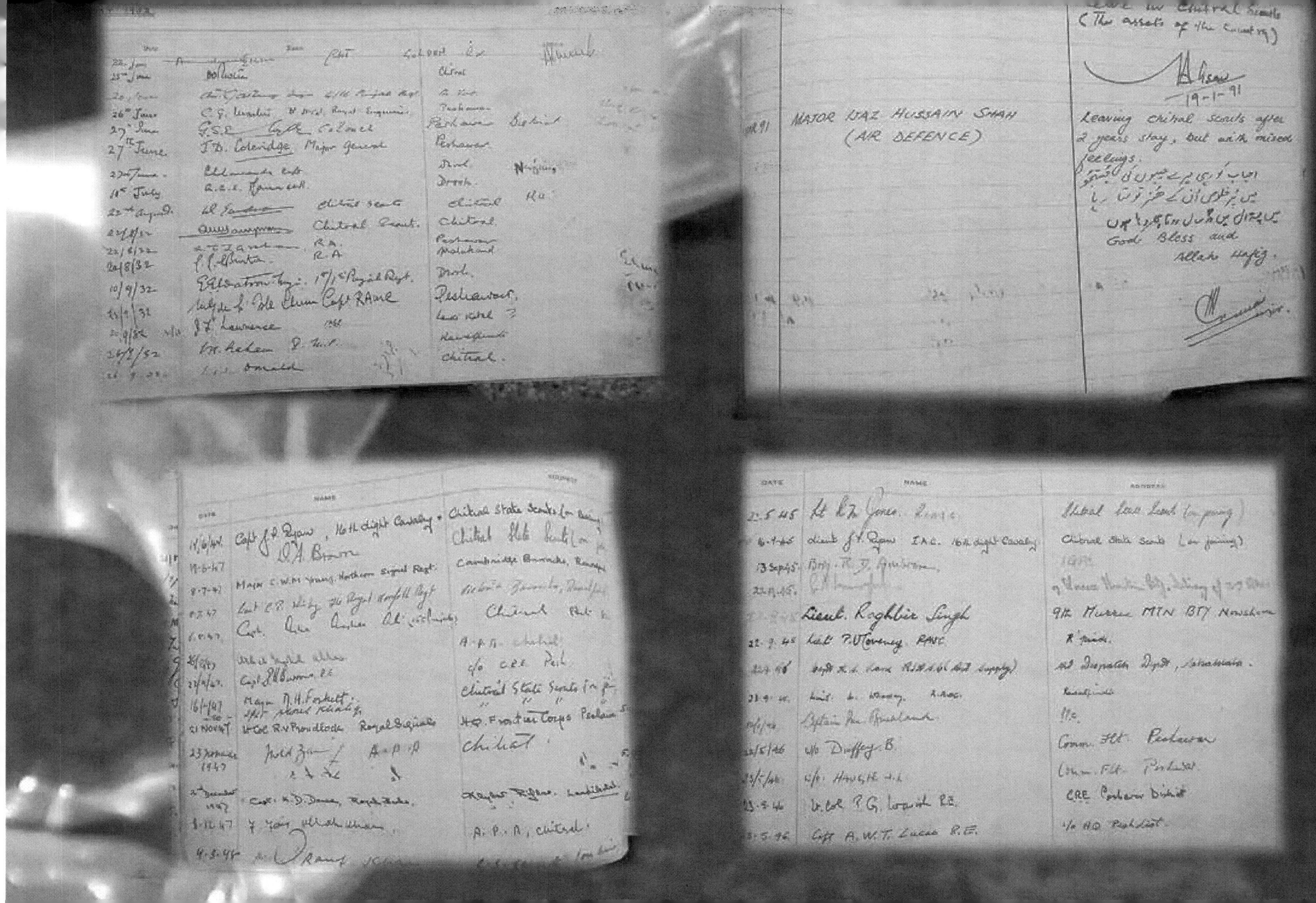

(The assets of the country)
19-1-91
MAJOR IJAZ HUSSAIN SHAH
(AIR DEFENCE)
Leaving Chitral Scouts after 2 years stay, but with mixed feelings.
God Bless and Allah Hafiz.
J.D. Coleridge Major General
Peshawar
Chitral Scouts
DATE
NAME
ADDRESS
Capt J.F. Ryan, 16th Light Cavalry
Chitral State Scouts
Major C.W.M. Young, Northern Signal Regt.
Lieut. Raghbir Singh
9th Murree MTN BTY Nowshera
Lieut. J.F. Ryan I.A.C. 16th Light Cavalry
Chitral State Scouts (on joining)
Lieut P.V. Coveney RASC
W/O Duffey B.
Comm. Flt. Peshawar
CRE Peshawar District
Capt A.W.T. Lucas R.E.
A.P.A. Chitral

Epilogue - History of History

History of Chitral Scouts is part of the project launched by Major General Ghayur Mahmood, Inspector General Frontier Corps, KPK , in May 2013, to compile the history of Frontier Corps. Existing material on Chitral Scouts is outdated as the last book written by an officer of Chitral Scouts was in 1972 'Chitral & Kafiristan a personal narrative' and due to then prevailing military culture it only dealt with the civil affairs and Chitral Scouts were not even mentioned. The first detailed account of Chitral appears in Major Robertson's two books, 'Kafiristan' and 'Chitral Story of Minor siege'. Rudyard Kipling made Chitral famous with his novel The Man who Would be King, later a movie was made on it starring Sean Connery and Michael Cain. 'Ghuzni, Kabul and Afghanistan' by G.T. Vigne in 1838 highlights the cultural clash between Kafirs and Muslims. Chitral does not name in any of the Mughul emperors autobiographies however the name Kafiristan strikes again and again. Thus Chitral to some extent has become synonymous with Kafiristan and this is where the real fame of Chitral lies. But there is more to Chitral than mere Kafirs, the Terich Mir peak is a living, geographical giant of Chitral's legendary past.

I first heard about Chitral Scouts in 1989 when Lieutenant Colonel Murad Nayyar committed suicide; I was a young captain then. In 1995 I first travelled through Chitral on a leave and that was my first interaction with Chitral Scouts, I stayed in the mess, it was October and Kalash valley was lively; I then trekked across Shandur with a night stay at Mastuj. Later in the years I was posted at Skardu army aviation squadron and had the opportunity to land at Chitral on few occasions. The flight path followed is from Gilgit, crossing Shandur and then following River Yarkhun; there were always anxious moments before Chitral airfield became visible. Colonel Awan was the commandant then and being old 'Air Defender' he was extra courteous to us pilots. Presently I visited Chitral Scouts first in Ramadan {June-July 2013} with my friend Major Kayani of army aviation and his civilian friends, aim was to get good photographs , however limitation is of handling so many photographers. We stayed at Garm Chashma and at Kalash in civil set up and this allowed an in-depth probing of culture. I lost my borrowed camera at Kalash and realized at night, however the hotel waiter himself a Kalashi went to the Bambouret Village, opened up all the shops which I have visited and found the camera under garments; this highlights the level of honesty among Kafirs.

Major Langland and Madam Carey Schoffield

It was in this first trip that I met Major {retired} Siraj Ul Mulk an ex army aviator and son of illustrated Colonel Khushwaqt ul Mulk who was among the pioneer Muslim officers of Pakistan Army and a veteran Frontier

44 https://www.thegazette.co.uk/London/issue/59446/supplement/B3.

Naik Wasi

Naik Nazir

Captain Buttar

Havildar Javaid

Naik Nazir's brother and father

Corps officer. Hindu Kush Heights is a hotel west of Chitral airport and city, its opening in 1999 was an international headline as no fewer than ten aircraft of all kind piloted by Siraj's friend flew across the globe to attend the ceremony; Siraj is a living history of Chitral's princely affairs. The other institution in Chitral is the Major Langland High School, Major Langland is now 95 years of age, he was born in 1917 , his father had died couple of weeks earlier in the war. Langland and his brother were raised by their grandparents. In 1939's September, on the evening of Britain declaring war, Langland then teaching mathematics in a school simply switched off the radio and went downstairs to get him self enrolled in army. Fate brought him to India and he stayed with Pakistan Army after independence, out of his sheer love. After his retirement in 1953 he was adjusted in Aitcheson College Lahore by General Ayub Khan, later in years Langland established Razmak Cadet College and in 1989 was persuaded by the deputy commissioner of Chitral to stay and establish a school here. Major Langland thus became a part of Chitral's legacy, he established a school for girls and boys and later expanded to a college at Sangor which is close to the Hindukush Heights; in all these years Pakistan Government{ Benazir Bhutto and Pervaiz Musharraf}, Army, and Chitral Scouts helped the project in own capacity. Major Langland despite his old age stayed as the principal of the school and took active classes as well, in 2010 he started searching for his successor and this is how the current principal Carey Schoffield a Cambridge and Oxford graduate and authoress of '*Inside Pakistan Army War against Terror, a woman's narrative*' volunteered herself. I had met Major Langland and been his host for three days

Major (R) Ali Imran Shah

at Kelly House at Aitcheson College last year thus I was keen to meet him and came to know about all this progress. I did not have the opportunity to meet the Commandant Colonel Jadoon as he was not feeling well; in Scouts Corps nothing moves without commandant's nod or approval. I still managed to gain some data thanks mainly to the Adjutant, Major Kazmi. I took the return journey to Peshawar on a public transport, it further helped in understanding the culture and peculiarities of the time.

Major (R) Ahmadul Mulk he is senior most living scout officer

My next visit to Chitral Scouts took place in September 2013, while I was at Khyber Rifles Mess, urgency was regarding a function at Major Langland School's Defence of Pakistan Day's celebration. It is at this particular point that the geographical and historical interconnection between Peshawar and Chitral becomes very obvious. Here I am at Khyber Pass and want to reach Chitral immediately and shortest way is through Kunar River banks taking me to Chitral by evening in all weather. But now I am compelled due to international treaties to travel back through Khyber Pass, Malakand Pass and cross Lowari at 10500 feet and then reach Chitral. In the end I was able to reach Chitral in twelve hours on a private taxi driven by an Afghan national settled in Malakand and thanks to modern roads, but this highlight the peculiar position of Chitral, in winter the Lowari used to be close down and all travelling was through Afghanistan but presently Lowari tunnel operates on few days a week in winter to facilitate the travelers. This highlights the importance of Lowari tunnel as well.

Major Tahir Kiani, Army Aviation

I stayed over a month with Chitral Scouts at Chitral, Drosh, Mirkani and Dommel Nisar, attending their functions like Bara Khana on 6th September at Chitral and farewell Bara Khana at Drosh of outgoing wing commander. I spent two full moon at Chitral, I was fortunate to stay at the village Sanoghar in the house of Naik Nazir and had a lively conversation with his father who retired as subedar major of 38 Frontier Corps regiment. A different kind of approach was adopted in gaining knowledge about the local history; through the essays of school children and write ups from the recruits. Chitral Scouts have one the finest library among all Frontier Corps corps, it is rich but the archives are not part of it, there is an 'old record room' inside the old Chitral Fort and also one at Drosh but they do not have record in order. I found the original report on the Chitral Foot

Bridge dating back to 1898 in the head clerk's office, hidden among a heap of loose papers. Present commandant took keen interest in the history and afforded me access to much talked about photo albums of Chitral Scouts, the fable about presence of Lieutenant Colonel Murad's diary is not correct. From the mess havildar's store amidst an antique almirah were found the visitors books, one more is located at the Quarter Guard of Chitral Scouts. These visitor books are important in tracking history, another important item were the 'Shikar Book' which were initiated in 1898 at Drosh and were maintained till 1931, an amazing record of military life of that era. Photographs are important part of history, one picture conveys thousand words dictum stands absolutely correct , Chitral Scouts somehow the other do not have pictorial record of their illustrious history as it should have been. There are no digital pictures beyond 2008 when majority of data was lost in a virus attack. Commandants pictures are placed outside in the Mess gallery and it resulted in their fading out, same is the story with wing commanders pictures. All these shortcomings have been overcome through an excellent cooperation of the Chitral Scouts. Wing commanders, wing subedars, education staff, mess staff and A&Q Branches of headquarters, all have facilitated in the compilation of history.

Chitral is on path of progress, although there is not much of difference in terms of roads when compare to 1995, yet the indicators of the development are visible. Mirkani- Arandu Road is under construction by FWO and so is the case with Mirkani-Drosh Road. Proper road starts almost twenty kilometers short of Chitral and terminates at the town itself. Track to Kafiristan is as primitive as mentioned by Lieutenant Colonel Afzal in 1972, Kafirs have modernized in many aspects, their women wears branded shoes but still they retain their cultural distinction. It is disturbing to see their privacy being so openly violated by

Major General Ghayur Mehmood, former Inspector General Frontier Corps

tourists with their cameras and video making equipment. The typical scene looks like as if tourist have entered into a living museum with exquisites humans objects which have to be photographed.

Chitral Scouts is the proverbial big brother of these Kafirs, it helps them in sustaining the life by providing security and jobs, arranging functions and projecting them in positive manner. Chitral is a walnut country, life revolves around walnut, its oil is used for cooking yet it is almost impossible to procure it from open market. Local delicacies are seldom available in bazaar and even rarely cooked in mess, apple and pear are in abundance but not used in local dishes, apple pie is simply unheard of in Chitral. Local transport is economical while commuting between the towns fare for public transport from Drosh –Madaglasht is 150 rupees one way and similar is the fare between Chitral –Booni ; on the other hand fare for taxi from Chitral-Peshawar is Rupees 9000/ and from Chitral Mess to Hindukush Heights is Rupees 700 only one way. Electricity crisis are severe in some parts of Chitral, majority of the area is free from this as they have hydel power generation, the project at Goleen is a living example of this. Petako Gaz, Drosh, Mirkani and Dommel Nisar all have diesel power generators which are operated on fixed time. Mobile phone services works in limited parts of Chitral, Pakistan telecommunication's wireless service is most popular, every scout {almost} have these sets which charge a fixed amount of rupees 320 for all month free calling inside Chitral. All villages have these wireless services also, there are no bicycles in Chitral, shops closed down at sunset, internet banking is available and so are the ATMs. Fruit is in abundance, it comes from adjoining valleys thus it differs in taste, Chitrali cloth is famous for making cap and headgear, most of the shops in the Shahi Bazaar deals with this. Winter is harsh and life pattern changes, city gets a blanket of snow in winter which cripples the life as transportation to valleys is discontinued, flights are cancelled and often snow slides wipes out the road infrastructure. But it has its own beauty and charm. I personally feel that Chitral is the most peaceful area in Pakistan to settle down.

Dr. Reena Rizvi.
She facilitated in having access to British Archives

Subedar Major (Retd)
Qilandar Shah

Major (Retd) Siraj-ul-Mulk in the center

Bibliography

1. Dough Vincent *A Buddhist Reading*, approved but unpublished Thesis, University of Western Ontario,2007 also see Buddha & His Doctrines, a bibliographical essay, Otto Kistner,{London,Trubnar & Company,1869}
2. James Cowles Prichard *Researches into the Physical History of Mankind*, Third Edition, Volume IV, [London 1844]
3. S.A. Malik *The Muslim Conquest of Central Asia*,{ Army Education Press, Rawalpindi, 1979}.
4. Aurel Stein, *The Geographical Journal* Volume,59, No.2 {February 1922} A Chinese expedition across the Pamirs and Hindukush, A.D. 747.
5. *Tuzk-i-Taimouri*, Autobiography of Tamerlane translated from Persian by Syed Abu al Hashim Nadvi {Sange Meel publishers, Lahore, 2012
6. *Tuzk i Babri*, Autobiography of Emperor Babur, translated from Persian into Urdu & English.
7. *NWFP, Imperial Gazetteer* 1905
8. MuhammadQaiserJanjua. *"In the Shadow of the Durand Line; Security, Stability, and the Future of Pakistan and Afghanistan"*. Naval Postgraduate School, Monterrey, California
9. Lieutenant Colonel Afzal Khan , Commandant Chitral Scouts 1971-1973. *Chitral & Kafiristan a personal study*{ Ferozsons, Lahore, 1975}.
10. Major Muhammad Nawaz {retired} *The Guardians of The Frontier, The Frontier Corps NWFP*, { Headquarters Frontier Corps, Peshawar, 1994},
11. Major George Robertson *Chitral Story of a Minor Siege*
12. Report of Captain McCardy 1930, Chitral Scouts Archive
13. Anila, Bali. The *Russo-Afghan Boundary Demarcation 1884-95: Britain and the Russian Threat to the Security of India*, (School of Humanities, University of Ulster. Thesis submitted for the degree of D.Phil.1985
14. J.A. MacGahan *Campaigning on The Oxus and the Fall of Khiva* (Simpson, London, 1874).
15. Military Report and Gazetteer of the Gilgit Agency and independent territories of Dariel and Tagir, {Government Monotype Press, Simla, 1903.
16. Frontier and Overseas Expeditions from India, volume I,
17. Munphool Pundit, *relationship between Gilgit, Chitral and Kashmir, Journal of the Ethnological Society of London*{1869-70}. Volume 2, No.I{1870}, http://www.jstor.org/stable/3014434, accessed on 24th January 2012.
18. *Gilgit The Northern Gates of India*, {Delhi, 1977}.

19. *Aithcheson, Treaties, Engagements and Sanads* 21

20. *Partition Of Punjab Official Papers.*

21. Alastair,Lamb, *Kashmir Disputed Legacy.*

22. Major William Brown, *The Gilgit Rebellion*, {Ibex, Rawalpindi, 1998].

23. Victoria Schofield *Kashmir in Conflict; India, Pakistan & the unending War* {London, Taurus, 2010} p-63.

24. Shuja Nawaz, *The First Kashmir War Revisited, India Review, Volume 7, No.2, April-June 2008*, pp.115-154. Taylor & Francis Group.

25. Yaqoob Khan Bangash, *'Three Forgotten Acessions, Gilgit,Hunza and Nagar', The Journal of Imperial and Commonwealth History"*, Volume38,No.1, March 2010.

27. Muhammad Yousaf , *Tareeq Baltisatn.*

28. Lieutenant Colonel Mirza Hasasn, *Shamsher say Zanjeer Tak*, urdu, autobiography, 1979,

29. Naik Iqrara Ud din Khusro, *Chitral aur Chitral Scouts Tareekh kaay ainey mein*, {urdu} unpublished, the writer is a naik in Chitral Scouts,

30. Captain James of Royal Engineers Report 1899, Chitral Scouts Archives, Chitral

31. Muhammad Rahim, Tareekh Injigan, (Hafiz, Chitral, 2006), pp, 16-58.

32. Munphool Pundit, *Chitral, Gilgit & Kashmir*, 1869, Also see G.T. Vigne, *A Personal Narrative of a visit to Ghuzni, Kabul, and Afghanistan*, first printed 1842, reprinted 1982, Sang-e-Meel, Lahore, p-180, p-385.

33. Dr G.W. Leitner *The Races and Languages of The Hindu Kush, Imperial and Asiatic Review. July-October 1891*. pp.139-156.

34. Frontier Corps Newsletter 1966, Chitral Scouts, page 31.

35. District Coordinating Officer Chitral, Letter No.4225 dated 1st October 2010.

37. DCO Chitral, Letter No.2818/DCO/dated 6th September 2011.

38. Interview with Naik Nasr Minullah, October 2013.

Shaheeds & Wounded. War Against Militancy

Shaheed

Ser	No , Rank & Name	Tribe	Date	Ser	No , Rank & Name	Tribe	Date
i.	403741 LNaik Shakir ud Din	Chitrali	05-1-2004	xix.	405358 Sepoy Mohsin Ali	Chitrali	Ni 7/8-1/08
ii.	406068 Sepoy Nasir Ullah Khan	Chitrali	05-1-2004	xx.	406130 Sepoy Mehrab ud Din	Chitrali	Ni 7/8-1/08
iii.	404291 LNaik Abdullah	Chitrali	02-4-2003	xxi.	403526 Havildar Buzurg Muhammad	Chitrali	07-5- 2008
iv.	403586 LNaik Muhammad Shafi	Chitrali	27-5-2005	xxii.	404624 LNaik Qadeer Khan	Chitrali	27-10-2008
v.	405494 Sepoy Sami Ud Din	Chitrali	27-5-2005	xxiii.	405429 Sepoy Haji Rehmat	Chitrali	6-11-2008
vi.	406382 LNaik Shuja Ul Islam	Chitrali	27-5-2005	xxiv.	407548 Sepoy Imtiaz Ahmad	Chitrali	6-11-2008
vii.	404552 LNaik Israr ud Din	Chitrali	23-4-2006	xxv.	405718 Sepoy Wali Khan	Chitrali	10-11-2008
viii.	404300 LNaik Karim ud Din	Chitrali	23-4-2006	xxvi.	403206 Havildar Jamshaid Ahmad	Chitrali	14-11-2008
ix.	405682 Sepoy Aziz ud Din	Chitrali	23-4-2006	xxvii.	801559 Naik Hamid Khan	Chitrali	02-1-2009
x.	403833 Sepoy Salah Ud Din	Chitrali	23-4-2006	xxviii.	803413 Sepoy Rahmat Gul	Chitrali	02-1-2009
xi.	403341 Sepoy Mumtaz Hussain	Chitrali	23-4-2006	xxix.	404002 Naik Jan Khan	Chitrali	03-2-2009
xii.	406913 Sepoy Walid Ahmad	Chitrali	23-4-2006	xxx.	401620 N/Subedar Nasir Ullah	Chitrali	27-4-2009
xiii.	404881 LNaik Saif Ullah	Chitrali	17-12-2006	xxxi.	405978 LNaik Bashir Ahmad	Chitrali	27-4-2009
xiv.	402833 Havildar Zaffar Shah	Chitrali	7/8-1/2008	xxxii.	405419 Sepoy Ajaib Khan	Chitrali	27-4-2009
xv.	403361 Havildar Muhammad Nayab Gul	Chitrali	Ni 7/8-1/08	xxxiii.	406115 Sepoy Bashir Ahmad	Chitrali	27-4-2009
xvi.	403333 Naik Mirza Gul	Chitrali	Ni 7/8-1/08	xxxiv.	406721 Sepoy Aman Ur Rehman	Chitrali	27-4-2009
xvii.	404402 Naik Mukhtar Ahmad	Chitrali	Ni 7/8-1/08	xxxv.	406834 Sepoy Sher Nawaz	Chitrali	27-4-2009
xviii.	405262 Cook/Sepoy Asraf Shah	Chitrali	Ni 7/8-1/08	xxxvi.	302939 Sepoy Zakir Ullah	Chitrali	27-4-2009

Ser	No , Rank & Name	Tribe	Date
xxxvii.	406496 Sepoy Sher Azam	Chitrali	27-4-2009
xxxviii.	406276 Sepoy Bahadar Shah	Chitrali	26-4-2009
xxxix.	406359 LNaik Shah Hussain	Chitrali	07 5-2009
xl.	404472 Naik Rehmat Jawan	Chitrali	26-5-2009
xli.	407011 Sepoy Zhanoyar	Chitrali	26-5-2009
xlii.	404000 Naik Mir Hakim Khan	Chitrali	04-6-2009
xliii.	406003 Sepoy Farzand Ali Shah	Chitrali	03-5-2009
xliv.	406361 SepoyShahzad Rahim	Chitrali	25-6-2009
xlv.	803588 Sepoy Gul Zahid	Chitrali	01-12-2009
xlvi.	400805 N/Subedar Zaffar Murad	Chitrali	14-1-2010
xlvii.	407615 Sepoy Fahim Ud Din	Chitrali	23-1-2010
xlviii.	407750 Sepoy Noor Ahmed	Chitrali	24-2-2010
xlix.	405557 Sepoy Sajjad Hussain	Chitrali	28-5-2010
l.	407306 Sepoy Javeed Iqbal	Chitrali	22-6-2010
li.	407387 Sepoy Sabir Ahmed	Chitrali	05- 2-2011
lii.	401495 Subedar Niat Ghazi	Chitrali	27-8-2011
liii.	401618 Subedar Younas Khan	Chitrali	27-8-2011
liv.	403712 Havildar Jalal Ud Din	Chitrali	27-8-2011
lv.	403998 Havildar Bajgi Khan	Chitrali	27-8-2011
lvi.	404076 Havildar Ali Muhammad	Chitrali	27-8-2011
lvii.	404278 Naik Nazir Ahmed	Chitrali	27-8-2011
lviii.	405379 Naik Manzoor Elahi	Chitrali	27-8-2011
lix.	404841 LNaik Rehmat Nizar Khan	Chitrali	27-8-2011
lx.	405449 LNaik Shahzar Wali Khan	Chitrali	27-8-2011
lxi.	495572 Sepoy Hazrat Ullah	Chitrali	27-8-2011
lxii.	406024 Sepoy Muhammad Shoaib	Chitrali	27-8-2011
lxiii.	406085 Sepoy Ali Haider	Chitrali	27-8-2011
lxiv.	406760 Sepoy Atta Ur Rehman	Chitrali	27-8-2011
lxv.	407476 Sepoy Sher Akbar	Chitrali	27-8-2011
lxvi.	407430 Sepoy Hazrat Umar	Chitrali	27-8-2011
lxvii.	407462 Sepoy Hashim Pannah	Chitrali	27-8-2011
lxviii.	803553 Sepoy Rehmat Sahib Khan	Chitrali	27-8-2011
lxix.	405440 Sepoy Ziaul Mulk	Chitrali	27-8-2011
lxx.	407944 Sepoy Tariq Mahmod	Chitrali	27-8-2011
lxxi.	407675 Sepoy Shali Khan	Chitrali	27-8-2011
lxxii.	404174 B/Sepoy Amir Zafar Khan	Chitrali	27-8-2011
lxxiii.	404736 C/Sepoy Sher Ahmed	Chitrali	27-8-2011
lxxiv.	405910 Sepoy Faqir Hussain	Chitrali	27-8-2011
lxxv.	403737 Naik Rehmat Akbar	Chitrali	16-10-2011

Injured

Ser	No , Rank & Name	Tribe	Date
i.	PA-32353 Major Ghulam Rasool	Jatt	26-4-2009 22-6-2010
ii.	400279 Subedar Muhammad Daud	Chitrali	16-9-2008
iii.	404968 LNaik Akbar Hayat	Chitrali	16-9-2008
iv.	407552 Sepoy Didar Ali Khan	Chitrali	16-9-2008
v.	407381 Sepoy Latif Ullah	Chitrali	16-9-2008
vi.	406362 Sepoy Burhan Ud Din	Chitrali	16-9-2008
vii.	407257 Sepoy Muhai Ud Din	Chitrali	16-9-2008
viii.	407118 Sepoy Muhamma Shakir Khan	Chitrali	16-9-2008
ix.	405626 Sepoy Miraj Ahmed	Chitrali	06-11-2008
x.	406985 Sepoy Mirza Ali Khan	Chitrali	06-11-2008
xi.	407542 Sepoy Kashif Ahmed	Chitrali	06-11-2008
xii.	407592 Sepoy Abdul Wali Khan	Chitrali	06-11-2008
xiii.	406012 Sepoy Ghulam Nabi Khan	Chitrali	06-11-2008
xiv.	407537 Sepoy Ikram Ullah	Chitrali	06-11-2008
xv.	405470 Sepoy Safir Ullah	Chitrali	06-11-2008
xvi.	40402 D/Sepoy Sultan Ur Rehman	Chitrali	06-11-2008
xvii.	405435 Sepoy Amir Hussain	Chitrali	14-11-2008
xviii.	405061 LNaik Sadiq Ur Rehman	Chitrali	18-11-2008
xix.	406573 LNaik Muhammad Karim Khan	Chitrali	25-01-2009
xx.	405942 Sepoy Ilyas Ahmed	Chitrali	04-5-2009
xxi.	403210 Havildar Sardar Hussain	Chitrali	05-5-2009
xxii.	406012 Sepoy Ghulam Nabi	Chitrali	07-6-2009
xxiii.	404394 Naik Muhammad Nadir	Chitrali	19-6-2009
xxiv.	404207 Naik Mubashir Hassan	Chitrali	27-6-2009
xxv.	407488 Sepoy Noor Wali Shah	Chitrali	27-6-2009
xxvi.	407276 Sepoy Salah Ud Din	Chitrali	29-10-2009
xxvii.	406943 Sepoy Muhsin Ali	Chitrali	08-5-2008
xxviii.	403489 Havildar Dildar Ali	Chitrali	19-9-2009
xxix.	404236 Naik Muhammad Kaim	Chitrali	19-9-2009
xxx.	406988 Sepoy Iqbal Ud Din	Chitrali	19-9-2009
xxxi.	407056 Sepoy Daulat Aman	Chitrali	19-9-2009
xxxii.	407304 Sepoy Ijaz Ali Shah	Chitrali	19-9-2009
xxxiii.	406438 Sepoy Sana Ullah	Chitrali	19-9-2009
xxxiv.	404993 Naik Sardar Wali	Chitrali	10-10-2010

Ser	No , Rank & Name	Tribe	Date	Ser	No , Rank & Name	Tribe	Date
xxxv.	404939 Naik Asad Ur Rehman	Chitrali	10-10-2010	liii.	405801 Sepoy Abdul Hafiz	Chitrali	26-4-2009
xxxvi.	403896 Naik Ali Nawaz	Chitrali	20-02-2009	liv.	402663 Sub Haider Wali Khan	Chitrali	30-8-2010
xxxvii.	404833 LNaik Aziz Ur Rehman	Chitrali	15-6-2009	lv.	404292 LNaik Hashim Ali	Chitrali	16-12-2009
xxxviii.	401097 N/Subedar Ehsan Ul haq	Chitrali	26-4-2009	lvi.	403816 Hav Mumtaz Ahmed	Chitrali	22-6-2010
xxxix.	403100 Havildar Muhammad Musa	Chitrali	26-5-2009	lvii.	404081 Naik Mehrab Hussain	Chitrali	22-6-2010
xl.	403291 Havildar Muhammad Fayaz	Chitrali	06-5-2009	lviii.	404459 Naik Sartaj Wali	Chitrali	22-6-2010
xli.	403338 Havildar Shakir Ullah	Chitrali	01-5-2009	lix.	404553 Naik Nasir Ullah	Chitrali	22-6-2010
xlii.	405164 LNaik Syed Kifayat Ullah	Chitrali	01-5-2009	lx.	404620 Naik Javed Iqbal	Chitrali	22-6-2010
xliii.	406449 LNaik Tika Khan	Chitrali	7-5-2009	lxi.	406312 LNaik Zar Wali	Chitrali	22-6-2010
xliv.	405094 LNaik Sahib Nizar	Chitrali	7-5-2009	lxii.	407009 Sepoy Shafi Ullah	Chitrali	22-6-2010
xlv.	405465 Sepoy Sher Wazir	Chitrali	01-5-2009	lxiii.	407190 Sepoy Dool Khan	Chitrali	22-6-2010
xlvi.	405534 Sepoy Israr Ud Din	Chitrali	06-5-2009	lxiv.	407199 Sepoy Shakir Illahi	Chitrali	22-6-2010
xlvii.	405563 Sepoy Siraj Ud Din	Chitrali	7-5-2009	lxv.	407220 Sepoy Atta Ur Rehman	Chitrali	22-6-2010
xlviii.	405772 Sepoy Abdul Rehman	Chitrali	26-4-2009	lxvii.	403816 Hav Mumtaz Ahmed	Chitrali	22-6-2010
xlix	802132 Sepoy Rehman Ullah	Ranizai	26-4-2009	lxviii.	407638 Sepoy Maqsood Khan	Chitrali	22-6-2010
l.	406212 Sepoy Sarwar Ali Khan	Chitrali	01-5-2009	lxix.	903481 Sepoy Irshad Ahmed	Chitrali	22-6-2010
li.	406430 Sepoy Rehmat Khan	Chitrali	26-4-2009	lxx.	407522 Sepoy Muhammad Zakir	Chitrali	15-7-2010
lii.	406880 Sepoy Syed Siraj Ud Din	Chitrali	26-4-2009	lxxi.	407197 Sepoy Jahan Zaib	Chitrali	18-7-2010

Ser	No , Rank & Name	Tribe	Date
lxxii.	404320 Naik Muhammad Islam	Chitrali	17-8-2010
lxxiii.	402663 Sub Haidar Wali	Chitrali	30-8-2010
lxxiv.	403881 Naik Sarwar Ahmed	Chitrali	24-9-2010
lxxv.	406214 Sepoy Jamil Ahmed	Chitrali	27-9-2010
lxxvi.	405627 Naik Rehmat Nawaz	Chitrali	09-2-2011
lxxvii.	407358 Sepoy Inayat Ullah	Chitrali	07-4-2011
lxxviii.	406703 Sepoy Zahir Ud Din	Chitrali	02-5-2011
lxxix.	406903 Sepoy Rehmat Ullah	Chitrali	06-5-2011
lxxx.	407570 Sepoy Abdul Hamid Khan	Chitrali	17-6-2011
lxxxi.	406937 Naik Niaz Ur Rehman	Chitrali	17-7-2011
lxxxii.	403289 N/Sub Pukhtoon Wali	Chitrali	27-8-2011
lxxxiv.	403356 N/Sub Karim Ali Khan	Chitrali	27-8-2011
lxxxv.	405070 Naik Muhammad Yamin	Chitrali	27-8-2011
lxxxvi.	405382 LNaik Nadir Muhammad	Chitrali	27-8-2011
lxxxvii.	406707 Sepoy Wali Amin Khan	Chitrali	27-8-2011
lxxxviii.	406946 Sepoy Muhammad Iqbal	Chitrali	27-8-2011
lxxxix.	407226 Sepoy Nasir Ul Amin	Chitrali	27-8-2011
xc.	407555 Sepoy Anwar Ul Haq	Chitrali	27-8-2011
xci.	407742 Sepoy Sultan Ud Din	Chitrali	27-8-2011
xcii.	407232 Sepoy Fazal Mabood	Chitrali	27-8-2011
xciii.	407806 Sepoy Hayat Muhammad	Chitrali	27-8-2011
xciv.	407945 Sepoy Khurshid Ali	Chitrali	27-8-2011
xcv.	406194 Sepoy Ghulam Muhammad	Chitrali	27-8-2011
xcvi.	407969 Sepoy Amir Muhammad	Chitrali	27-8-2011

Author

Major Aamir Mushtaq Cheema was commissioned in 1984 {70th Long Course} in 58 Medium Ack Ack Regiment, he joined Army Aviation in 1989 {P-30} and retired in 2008. He holds a Master degree in history {Allama Iqbal Open University}, Master of Letters in War Studies {University of Glasgow}, presently he is a scholar of Ph.D in International Affairs in National Defence University, Islamabad. He is author of 'History of Royal Air Observation Corps 1936-1956', 'North Waziristan Militia & Tochi Scouts 1895-2013'; He also co-authored 'History of Pakistan Army Aviation 1947-2007'

Author with the Kafir toddler in Kalash Valley

CHITRAL SCOUTS

Chitral Scouts Mascot
8 May 1978 - 10 December 1998

Made in United States
Orlando, FL
04 November 2026

90207196R00101